AMERICA'S FOUNDING FRUIT

The Cranberry in a New Environment

SUSAN PLAYFAIR

UNIVERSITY PRESS OF NEW ENGLAND
Hanover and London

University Press of New England
www.upne.com
© 2014 University Press of New England
All rights reserved
Manufactured in the United States of America
Designed by Vicki Kuskowski
Typeset in Granjon by Copperline Book Services, Inc.

Library of Congress Cataloging-in-Publication Data
Playfair, Susan R.
America's founding fruit : the cranberry in a new environment / Susan Playfair.
    pages cm.
Includes index.
ISBN 978-1-61168-631-9 (cloth : alk. paper)—ISBN 978-1-61168-632-6
(pbk. : alk. paper)—ISBN 978-1-61168-633-3 (ebook)
1. Cranberries—United States. 2. Cranberries—United States—History.
I. Title. II. Title: Cranberry in a new environment.
SB383.P53 2014
634'.760973—dc23 2014013667

5 4 3 2 1

Illustration for the introduction, chapters 1–14, and the epilogue, ©2013 Ji Liang.
Illustration for "For the Cook (Recipes)," "Drawing with Bowl," from an original illustration by Holly Hurd DiMauro, Mossy Pond Prints.

CONTENTS

 INTRODUCTION

Without a farming population, a nation is never healthy in spirit.
It strays into the landscape of illusion and climbs the hill of dreams.
Yet only through an awareness of things as they are may we
Raise a humane and tolerant order.

—Henry Beston (Sheahan), "The Great Realities," 2000

Our seasons are defined by the biological rhythms around us. We count the days until we can pick fresh blueberries in June, gather blackberries in July, catch striped bass and dig clams in August, swallow oysters in September, and harvest cranberries in October. But something is changing.

On Sunday, September 16, 2012, satellite images of Earth indicated the greatest melting of the Arctic's summer sea ice recorded since 1979, the year when it first became possible to measure the degree of seasonal melt. The ice is melting because the land and sea around us are gradually warming and the plants, birds, and fish are either having to adapt or disappear.

This book is about one native fruit, the cranberry, how it is adapting, and what could be lost if it can still be grown in other countries, but not ours. I have chosen to write about the cultivation of a particularly American fruit; the men and women who grow and harvest it; how it is grown; its history, development, and health benefits; and how man and nature may be affecting the berry's long-term survival.

The cranberry is a metaphor for agriculture and a lesson to us about our interdependency with nature. The timing of its flowering is a cautionary reminder of the condition Rachel Carson imagined fifty years ago when she wrote "a strange blight crept over the area and everything began to change." At the time she was referring to chemicals used for insect control, but the warning is as relevant today as then. Today's blight is CO_2. The change is a gradual warming of the planet and the adjustments that plants, birds, fish, and ultimately humans need to make to survive.

Demographic data indicates that in 2012 more people worked as computer software engineers than as farmers in the United States. New apps may make our lives easier, but to survive we need to eat.

My great-grandfather owned and managed cranberry bogs, and I recall "walking the bogs" with him when I was three years old. Since then, in researching this book, I have walked bogs from Nantucket Island, Massachusetts, to Coos Bay, Oregon, and from Tomah, Wisconsin, to the Pine Barrens of New Jersey. I loved every minute of it. I hope this book helps introduce you to some of the people I met along the way and to a piece of our American heritage that we may be in danger of losing.

When I first began to gather material for this book, talking with the various men and women who grow and harvest cranberries, I asked if they were harvesting earlier as a result of a gradually warming planet. With few exceptions, each farmer replied that the berries have always ripened at the same time. Usually, the answer was that nothing had changed in the past hundred years, and that the harvest, depending on the variety, was always from the end of September into the beginning of November, except on the West Coast where cranberries enjoy a later and longer growing season. Only after completing much of the writing did I discover that, despite the farmers' insistence that nothing has changed, I may have phrased the question incorrectly. I should have been asking if the plants flowered earlier in an especially warm year or, conversely, later in a cold year. The period when the plant is in flower is when the arrival of pollinators and other insects must be synchronized for both the pollinators and berry production. Flowering time for cranberries could shift the balance of nature in ways greater than just the survival of our iconic cranberry.

I was fortunate to meet two men who have kept records of cranberry flowering times for more than thirty years. Thanks to them, together with Richard Primack, Libby Ellwood, and Caroline Polgar of Boston University, I have been able to explore whether or not the cranberry might be flowering earlier, what that means for other species, and what might change when cranberry growers and the cranberry itself try to adapt to a changing climate.

On one level this book is about how farmers in a particular industry work. As readers, we can't very well understand the significance of losing the cranberry to other climates unless we understand how it is grown and who grows it. If growing conditions for a particular plant change, that plant can be grown somewhere else. For a single-crop farmer whose family may have grown cranberries or some other fruit in this country for four or five generations, the move to another country would not be easy and the loss of a native fruit that has been cultivated in this country for close to two hundred years would diminish our sense of who we are.

The cranberry industry may use the latest scanners, probes, and tracking devices to ensure the quality and safety of its products. Its members may employ windmills, solar panels, and computerized monitoring for efficiency and reduced energy consumption. It may sell its berries around the world, but it is still made up of a group of farmers dependent on the whims of nature; regulated by various local, state, and federal rulings; and controlled by the laws of supply and demand. As Tom Larrabee Sr., a grower on Nantucket Island, says as he looks out at Sankaty Head Lighthouse, "There is nowhere else on earth where you could possibly even dream of having this kind of job." Unless cranberry growers and the larger population make some significant changes, the dream may not be a reality for future generations.

AMERICA'S
FOUNDING
FRUIT

1

A PERFECT DESIGN

*We proceeded to Cranberry Lake, so called from the great quanti-
ties of cranberries growing in the swamps . . . On this lake there
are about one hundred and fifty good hunters. . . . This was one
inducement for settling here which was increased by the prospect of
a plentiful supply of fish, rice and cranberries, . . . winter comforts
of too great consequence to be slighted.*

—John Long, Indian interpreter and trader,
Voyages and Travels, 1791

In North America, if you drive past the Mayflower Lanes, Whip-
poorwill Paths, Beaver Dam Roads, and Honeysuckle Ways that
have been created to provide frontage for newly built homes in Massachusetts's Plymouth County, Wisconsin's Wood County, New Jersey's Suffolk County, or Washington's Skagit County, you may spot
a sandy road leading through underbrush. Follow it and you could
find yourself in a world dedicated to growing and harvesting a small,
red berry with a big role in the history and culture of the United
States.

This is the little berry that kept both Native Americans and Pilgrims alive through the winter months, was considered "the choicest product of the colony" when presented to King Charles in 1677,
served as Benjamin Franklin's cure for homesickness in 1769, pro-

tected sailors on whaling ships from scurvy, was served to General Grant's men in 1864, provided over a million pounds per year of fruit to our World War II doughboys, has been a staple served at countless Thanksgivings, and is currently warding off various cancers. This is America's superfruit.

Early settlers called it a "craneberry" because its flower, angled head on upright stem, reminded them of the cranes who shared the earth and sky with them. We call it a cranberry. It is one of only three cultivated fruit native to North America. Blueberries and Concord grapes are the others. Partly because of its longer shelf life and thick waxy skin, the cranberry has the most flexibility. It can be boiled in sauces, baked in a pie, frozen in sorbet, dried in cereals or salads, pulverized in capsules, and even served smoked. A fresh cranberry can be taken out of the freezer after three years and added to your favorite muffin recipe, with no loss to the flavor or texture.

For me, cranberries always have meant the Federal Furnace Road in Plymouth, Massachusetts. Whenever I drive beneath its allée of fir trees, my anticipation begins to increase, and I wonder if the vines have changed color, if the bogs have been flooded, if the sprinklers have begun spraying water to ward off frost. Above all, I want to feel that sense of assurance that the four seasons still lend themselves to the growth of a small, red berry much as they did in the days when my great-grandparents farmed cranberries, as did their parents and grandparents. I, like so many others, seek the permanency and continuity that cranberry bogs have afforded for so long.

Interlocking branches form a canopy overhead to funnel me through the woods. Suddenly they separate to reveal an azure sky and a cranberry bog ripe for harvesting, horizontal sheets of deep red stretching before me. To my right is a graying wood screen house, as much a part of local history as the bleached bones of a prehistoric mammal are to our larger history. These buildings were where the cranberries were separated by hand and packed in wooden boxes for shipment, before Ocean Spray and the larger independent handlers took over the job of processing. Downhill from the screen house, Piney Woods, one of the oldest cranberry bogs still being harvested, nestles beneath hemlock and stately pines, its three acres just fitting the "S" curve of the road.

The temperature reads seventy-three degrees. Morning clouds have disappeared to allow the sun to highlight a wash of red berries on indigo blue water. The berries are Early Blacks, the first variety to ripen each fall, and Piney Woods is one of the first cranberry bogs to be harvested this year.

Before the sun rose above the trees, Rob Beaton and his foreman, Israel Cordero, flooded the bog and drove partly submerged water-reel harvesters, known as "water beaters," across it to loosen the ripe berries and allow them to float to the surface. Now Kenny Higgins, one of the Beaton crew, pulls a black line of plastic through the water to corral the berries and guide them toward the Beaton family's new pump machine. Kenny's short, wiry build in a wet suit visually extends the floating black line. At the far end of the bog, Mak Song and May Lee, two seasonal workers originally from Cambodia, rake berries into the corral and away from the edge.

Since the berry was first cultivated in 1816, cranberry bogs have provided a means for waves of immigrants to get a foothold in this country. Italians followed Finns. Portuguese from Cape Verde followed the Italians. Frederick Swift, a Cape Cod bog owner, wrote in his journal on Tuesday, April 10, 1900, "Four or five Portuguese came today & I put them in Parker House." The following day's entry reads, "Russ, Charlie & four Portuguese & FKS [Frederick K. Swift] cut 35 bags [of] vines today. I made 10 hrs."

The earliest cultivators of cranberries were the settlers who had emigrated from England, in some cases via Holland. Most of the nineteenth- and twentieth-century immigrants employed by the new bog owners had been farmers in their own countries. They had come from rural towns, where they understood hard physical labor and were willing to trade that for the chance to work outside and, at least temporarily, move out of their crowded tenements in Boston, Lawrence, Brockton, and New Bedford, the mill towns that claimed so many who had moved to the cities. They and their families were given "bog shanties" to live in during the best months of the year, from planting to harvest. And the pay was good. For the bog owners, recent immigrants were willing to work hard, share accommodations, and provide cheap, needed labor. At the end of the season, they were also expendable.

Thanks to the Resettlement and Reintegration Program, an international effort whose objectives include poverty alleviation and the regeneration of civil society, Mak, May, and some of their fellow Cambodian farmers are part of the latest ethnic group to work the American cranberry bogs. As the men gradually move the floating berries into an ever-smaller circle, their laughter and the lilt of a song from their birth country float across to me from the other side of the bog.

Mating dragonflies cavort overhead. Frogs squeak as they leap from their recently submerged homes into the water or skitter across a sea of berries. Passing drivers park along the edge of the road and exit their cars, cameras in hand, to record a symbolic cranberry harvest on the first day of fall. They raise their eyes from viewfinders long enough to register a blue, sixteen-wheel transport truck backing down a sandy incline alongside the bog's pump machine.

After positioning the truck's bed under the machine's overhanging chute, Bruce Taylor jumps down from the cab and looks at the red circle of cranberries corralled on the Piney Woods bog. He places his bet, an annual wager with Israel, at 350 barrels (35,000 pounds).

Israel, "Papo" to his crew, looks wistfully at the results of his efforts for the past year. His lips compress. His message: "maybe." He shrugs his shoulders, raising the upper half of the one-piece wet suit he is wearing. "These vines, they're close to a hundred years old. They've given us a lot of berries over the years."

Kenny drags a metal funnel and hose through the water, finally placing the funnel at the spot where it can best deliver the berries to the pump machine. When the optimum position has been established, Mak and another man pound in four metal stakes, one at each corner of the funnel to indicate its location below the surface. After checking their placement of the funnel, Israel ascends blue metal scaffolding to the controls for the new machine, then pulls several shiny levers. With a gushing sound, berries begin to be sucked in between the four stakes, then spit out at the top of the machine where spigots release a spray of water to clean them. The good berries slide down a grated chute into the front end of Bruce's truck box. Weeds and other debris are funneled into a small pickup truck ahead of the machine. A second hose spews clean, filtered water back to the bog.

The new machine cost close to $200,000. "But," Israel tells me, "it will pay for itself in four or five years, because Ocean Spray pays more for clean berries." He then climbs higher to the top of the scaffolding, where he cleans extra weeds away from the water spigots and checks the level of the berries in the truck. Berries are now mounded higher than the truck's sides. Israel has judged the timing so that not one berry has spilled over. From his perch, he motions for Bruce to move the truck forward, allowing the rest of the berries to flow to the back of the truck.

On the bog, Mac slowly pulls the plastic boom to tighten the diminishing circle of red while Kenny continues to rake berries toward the suction point. The circle of berries is shrinking as the blue of the water is growing. An apprehensive Bruce looks at the red shape, then back toward the truck where berries continue to spew into the box. His earlier bet is beginning to look on the low side. As Bruce ponders his estimate, a light breeze riffles the surface of the flooded bog and sends a sweet brackish smell our way.

The last berries are siphoned off the bog and the truck is completely full. Bruce starts the engine and backs out, heading for Ocean Spray's satellite processing center, a few miles down the road. He is transporting one of the earliest harvests. And he picked the right day. There is only one other truck ahead of him in line. "Tomorrow, next day," Bruce tells me, "trucks will be three deep. You can wait an hour or two in line here." He drives the truck onto a weighing scale built into the road and jumps out to handle the paperwork.

Inside the delivery station, Joe Dutro, the manager, sits at a window scanning a computer displaying the contents of Bruce's truck. Outside, scaffolding leads to the spotters' vantage point, where a man and woman check for the color, cleanliness, size, and condition of the berries in the truck before determining their value.

Whoosh! A hydraulic probe appears and insinuates itself into the mass of berries, where it sucks up random samples, then deposits them into white buckets to be further scrutinized. A second probe dives into a different section of berries as two more trucks pull into the queue.

The load weighed and sampled, Bruce heads up the hill to the processing plant to wait for a go-ahead from an Ocean Spray employee. High above the truck, elevated conveyors transport berries to be washed and cleaned. Then an arm reaches out a window to display a card with "No. 2" written on it. In our digital age, this act seems like a time warp.

Bruce pulls forward and prepares to back into the No. 2 aisle. After expertly backing the truck straight down the middle lane of three, he hops out to disengage what he refers to as the "landing gear" before covering the remaining distance. Berries slide out the raised back of the truck and down into a hopper to be fed onto one of the metal elevated conveyors, then moved to another bath and transported inside, where they will be inspected, sorted, bagged, and labeled or processed for juice.

From the moment we left the queue at the entrance, the berries in Bruce's truck were systematically cataloged with the time and date and the bog where they originated. Each bottle of juice or bag of fresh berries available to the consumer from a grocery store shelf can be tracked to this one particular batch of berries from this one particular cranberry bog. In the event of any tainted item bearing an Ocean Spray label, the source of contamination can be easily traced.

Bruce pulls a lever to lower the truck box, then climbs back up into the cab. At the loading station, he parks the now-empty truck on the exit scale. Then he goes inside to get the paperwork and to find out how close his initial estimate was. The final count: 44,160 pounds, or 441 to 442 barrels. For a year where it rained throughout all of June and most of July, for a year dependent on persnickety honeybees who don't work in the rain, and for a hundred-year-old bog, that's not a bad yield. Israel will be pleased.

Piney Woods's owners are the only cranberry growers Bruce works for. After checking back with Israel, he maneuvers the truck up the sandy incline to where the edge of the bog meets Federal Furnace Road. Then he turns the wheel to head south to another Beaton-owned bog to pick up a second load of berries and start the process again.

This is the traditional harvest for nonorganic berries, those grown with chemical pest management and destined to be made into juice or sweetened dried cranberries. It is the least labor-intensive, the most efficient, and arguably the most aesthetic method of harvesting cranberries today. It is also a far cry from the days when twenty or so men, women, and children in straw sun hats, canvas aprons, knee patches, and cotton gloves slowly moved across a bog on their hands and knees to gather the berries in long-tined wooden scoops.

With several variations, the harvest process is being replicated in the Massachusetts bogs of Nantucket, Martha's Vineyard, Wareham, and Plymouth; in the Wisconsin bogs of Eagle River, Tomah, and Warrens; in the Washington bogs of Greyland and Long Beach; in the bogs of Port Orford, Oregon; and wherever the winters are cold enough, the summers long enough, and the climate conditions predictable enough to sustain growth of the berry.

The cranberry's story began as the glaciers retreated, about fifteen thousand years ago, and chunks of moving ice scoured out depressions, then remained behind to melt and form lakes and ponds. Over time, moss and other vegetation began to grow around the edge of the "kettle basins." Wind blew leaves from the newly formed shrubs and bushes into the water. There they decayed to form a dense sponge that held moisture through the warm summer months. Deposits left by the glaciers morphed into a form of limonite or hydrated iron oxide. The wind and rain carried small amounts of this iron-rich sand down from the hills and over the winter ice. Little by little, the pond and its upland hills of sand became the ideal garden for a ground vine and its deep-red berry.

The cranberry, *Vaccinium macrocarpon* Alt is a perennial vine, native to Massachusetts, parts of Rhode Island and Maine, and Canada. Its smaller, far-removed cousin, *Vaccinium oxycoccos* L., grows wild in limited quantities within the United States in Oregon, Washington, Alaska, New Jersey, Pennsylvania, Indiana, Wisconsin, and Minnesota. It also grows in Russia, Germany, Finland, the British Isles, and much of Europe. The larger Massachusetts berry can be found growing on Holland's Terschelling Island, not as a native, but as a descendant of the North American berries that floated ashore from a seventeenth-century shipwreck. Legend has it that a barrel was discovered on the beach by a Dutchman, Jan Sipkes Cupido, who is reported to have saved the barrel and thrown the berries in the sand. Then the tides carried those seemingly worthless berries into the lowlands, where they subsequently flourished.

In the eastern United States the average precipitation has ranged from forty to eighty inches per year. Summers are short and winters long, thus allowing for the colder temperatures followed by warm months required for cranberries to set, then break dormancy, begin to flower, and ultimately to produce berries.

American Indian ancestors between the ice age and the later agricultural revolution had learned to preserve meat from the animals they trapped by drying it, cutting it into strips, pounding it, mashing it with suet and cranberries, and molding it into rounded, flattened balls. Later generations invented blowguns and bows and arrows to hunt but clung to the same means of preserving the meat and berries. The Wampanoag and Aquinnahs called the berry *sassamanesh*, or "sour berry," for its tart taste. In late summer and fall, when the berries were ripe, they picked them and stored them to be used throughout the winter, often mixing them with corn meal and water for a hearty soup. More nomadic tribes dried them whole or ground them and mixed them with fat and meat from deer, bear, or moose before pounding or molding them into small cakes, or *pimmegan*, that they stored for lean winter months.

The Pequod of Cape Cod and the Leni-Lenape of New Jersey called the berry *ibimi*, meaning "bitter berry." Both tribes are reputed to have roasted fresh whole cranberries, then applied them as a poultice to wounds suffered from poisoned arrows, thus drawing out the venom. According to the few remaining letters, journals, and diaries, the Wampanoag made a sauce of cranberries that they fed to patients believed to suffer from "cancers."

When English settlers arrived on Plymouth shores, they brought with them palates that craved the gooseberries they used in sauces for meats, filling for tarts, and the sweet fruit for puddings with names like "grunt," "duff," and "spotted dog." They found few native gooseberries, but they noticed bright red berries that grew close to the ground and were ripe for picking from September into November. But, oh how those berries puckered their mouths and were bitter to the taste. Eventually, the pilgrims began to sweeten the berries with maple syrup, as the Mohicans and other inland tribes did. They also learned from the natives how to prepare "craneberry tea" as a diuretic and to ward off urinary tract infections.

Ironically, the *Oxycoccus palustris*, a smaller cousin to the so-called American cranberry, or *Oxycoccus macrocarpus*, was abundant in English and Irish fens, but its acrid taste and flavor of tannin reserved its use primarily for jams and jellies, where it could be liberally sweetened. It was the size of a small pea and a light, almost pink shade, so it's conceivable that the early settlers from England didn't at first make a connection to the more robust, waxy, red berry they found in their new world.

In 1633 an Englishman, John Josselyn, sailed to Boston to visit his only brother and subsequently wrote *New-Englands Rarities Discovered*, a book on the flora and fauna of "thofe remote parts of the World." Published upon his return in 1672, it lists his discovery of the cranberry and some of its health benefits:

Cran Berry, or Bear Berry, becaufe Bears ufe much to feed upon them, if a fmall trayling Plant that grows in Salt Marfhes that are over-grown with Mofs; the tender Branches (which are red-difh) run out in great length, lying flat on the groun, where at diftances, they take Root, over-fpreading fometimes half a fcore Acres, fometimes in fmall patches of aboot a Rood or the like; the Leaves are like Box, but greener, thick and gliftering; the Blossoms are very like our English Night Shade, after which fucceed the Berries, hanging by long fmall foot ftalks, no bigger than a hair; at firft they are of a pale yellow Colour, afterwards red, and as big as a Cherry; fome perfectly round, others Oval, all of them hollow, of a fower aftringent tafte; they are ripe in August and September.

FOR THE SCURVY

They are excellent againft the Scurvy.

FOR THE HEAT IN FEAVERS

They are alfo good to allay the fervour of hot Difeafes.

We know from the Lewis and Clark journals that the men knew about the tart, red berry. In the diary entry for December 10, 1805, Joseph Whitehouse, a member of the expedition, wrote,

> The land betwen this and the Ocean is covered with Pine Trees, and on the Coast, low flatt land considerable Prairies and some swamps, in which grows Cranberries.

Members of the expedition who established Fort Clatsop noticed that the natives mixed the tart, red berries with dried meat, probably to make a kind of pemmican, similar to the recipe of the tribes to the east. The women used the juice for dyeing rugs, blankets, and cloth, while the men used the berry as a poultice on arrow wounds and as a cure for indigestion.

By February 1806 Captains Lewis and Clark both exhibited an awareness of the curative powers of cranberries. Possibly they had learned this from watching the native Clatsops, or they may have brought the knowledge with them from the East Coast. In any case, they both make mention in their journals that Collins, a member of the expedition, returned from a hunt with "Some Cranberries for the Sick." After that, both men describe the various illnesses suffered by members of the expedition . . . Bratten's "obvious cough and pain in his back" and Willard's "high fever" and "pain in his head and want of appetite." Whether or not they believed the cranberries were able to cure Bratten's and Willard's maladies is unclear, but they clearly believed that the cranberry was useful for its medicinal properties.

Scurvy, resulting from a lack of fresh fruit or Vitamin C, was an often fatal disease for seamen. The symptoms included a general lassitude due to capillary weakness that manifested itself in boils, swollen legs, loose teeth, and swollen joints. Until 1844, when it became mandatory for a seaman to swallow a prescribed amount of lime juice, the disease was weakening the British navy.

Sailors to the Americas noted as early as the sixteenth century that the Wampanoags, Algonquins, and other native tribes who ate cranberries as part of their diets didn't get scurvy. By the mid-1800s no self-respecting captain set sail from an American port without barrels of cranberries aboard. The fruit's waxy coating retarded spoilage, and cranberries were easier to get than the limes carried on English vessels. To ward off scurvy, a ship's purser doled out one handful of cranberries a day to each seaman on board.

From 1846 to 1873 Capt. John Hopkins Baker of Westport, Massachusetts, owned a series of fishing schooners. He also owned cranberry bogs. According to the bills of lading, on October 22, 1870, he shipped twenty-six barrels of cranberries to his vessel, the *Mary Chapin*, docked at Central Wharf in New Bedford, Massachusetts. The ship's account was billed $5.20. On November 2, 1870, a shipment was delivered to the wharf for twenty barrels, billed at $4.00. Three barrels were shipped to the wharf on November 9 for $0.75, and on November 30, two hundred barrels were delivered and billed at $50.00. The last three deliveries may have been to sailing ships other than the *Mary Chapin*, possibly to whalers. Clearly, the last delivery was to a ship planning a longer voyage, as was usually the case with whalers.

The wife of an officer stationed in the Wisconsin region sent a letter home attesting to some of the medicinal uses Native Americans in the area subscribed to cranberries:

> The Indians are the cranberry gatherers. [They] attribute great medicinal virtues to the cranberry, either cooked or raw; in the uncooked state the berry is harsh and very astringent; they use it in dysentery, and also in applications as a poultice to wounds, and inflammatory tumours, with great effect.

Fur traders in the area seeking cranberries for the fruit's medicinal value traded beads, knives, blankets, and whiskey with the Chippewa, Winnebago, Menominee, and Potawatomi. Sadly, after 1850, under the "swampsland grant," the tribes were forced to give most of their land to the U.S. government, along with their right to pick the berries and trade them.

Less is known about the ways Native Americans in the Washington and Oregon areas used cranberries. "The Western tribal story is sad," says Bill Snyder, former director of economic development for the Coquille Indian tribe in Coos Bay, Oregon. "After first contact with outsiders, in the 1850s, it didn't take long for the tribes to be decimated with disease and hostilities. Tribal members who survived were relocated to reservations, and most medicinal uses for cranberries and other indigenous plants were lost in the process."

Over seventy-five years later, when scientists had begun to record oral stories from the remaining natives, Huron Smith, an ethnobotanist, collected plants in the area of the Potawatomi tribe in upper Minnesota. He then showed his specimens to various tribal members, asking them to identify the plants using the native name and to explain how they used them. The forest-dwelling Potawatomi claimed not to use the cranberry for medicinal purposes, but then added, "all of [our] native foods are also at the same time medicines and will maintain health." In explanation, Smith added that he was certain the Potawatomi knew what was in different plants and herbs they used to treat various maladies and also which ones to use for different illnesses, but that their medicinal practices were sacred and thus should not be divulged.

Today, scientific findings are corroborating many of the medicinal practices of Native Americans, early traders and explorers, and nineteenth-century ship captains. Any woman who has once suffered from a urinary tract infection has been told to drink cranberry juice to ward off future attacks. Less well known is the cellular research being conducted by various research scientists in Canada and the United States.

Since the year 2000 Dr. Catherine Neto, professor of chemistry at UMass Dartmouth, has published or been part of a team of researchers on at least ten papers involving studies on cranberries and human health. Dr. Neto has the focused eyes of a research scientist and the relaxed warmth of a mother of two college-age children. As she guides me through her lab, I ask if she regularly drinks cranberry juice.

"Absolutely!"

"And you eat dried cranberries?"

"But of course."

When asked how she had decided to focus her studies on cranberries, she explains that she had been aware of the curative reputation cranberries enjoyed, and she had read about the health benefits of wine and grapes. Knowing that cranberries have similar compounds to those found in grapes, she decided to look into the connection.

A polyphenol compound found in the red skin of grapes and cranberries, has the ability to induce cancer cells to undergo apoptosis, or disintegration, thereby enabling the harmful cells to be engulfed and absorbed by white blood cells and eliminated as waste. Sweetened dried cranberries, made up of the berry's skin and seeds and available to us in our breakfast cereals, trail mix, and salads provide the most undiluted form of resveratrol other than powder or pills, neither being as palatable or as colorful as the dried berry. The omega-3 in the seeds may also help keep our joints supple.

Based on what is known about the properties of polyphenols, Dr. Neto and her graduate students study the effects of the proanthocyanidins or PACs, a type of polyphenol found in cranberries. They are curious about the ability of these PACs to inhibit the growth of cancerous tumors in leukemia cells in the breast, lung, colon, and prostate. A paper Dr. Neto coauthored and published in the *Journal of the Science of Food and Agriculture* reports her findings that the structure of PACs in cranberries significantly inhibits the proliferation and growth of tumor cells.

Most fruits contain proanthocyanidins, but the PACs in cranberries are unique in the way they are linked and are thus believed to be more effective than the PACs in commercial apple juices, green tea, and dark chocolate as antiadhesion agents in blocking urinary tract infections from taking hold.

Native Americans believed that when members of the tribe ate cranberries as part of their regular diet, they were not only protected against urinary tract infections, tumors, ulcers, dysentery, scurvy, and dropsy, but they also kept their teeth healthy into old age. Perhaps this is the basis for the maxim "you are only as healthy as your teeth." Dr. Neto and other scientists attribute the promotion of healthy teeth to three ways that the cranberry PACs block bacterial adhesion: by changing the shape of the bacteria from rods to spheres, by altering the cell membranes of the bacteria, and by creating a barrier between the bacteria and cell walls. This combination tends to prohibit bacteria from adhering to otherwise healthy cells.

A pilot study conducted at Boston's Brigham and Women's Hospital in 2005 further clarified the cranberry's unique properties as a bacteria blocker by also studying the effect of raisins to inhibit the adherence of *E. coli* bacteria from attaching to cell walls. The results indicated that antiadhesion was exhibited in patients who had eaten sweetened dried cranberries but not in patients who had consumed raisins.

Before leaving her lab, I ask Dr. Neto if she has any advice for those of us who look to the cranberry as our preferred therapeutic fruit. She laughs, purses her lips in thought, and answers, "Well, if you think someone is having a stroke, give them a glass of cranberry juice." She then glances at me out of the corner of her eye to let me know that's what she would do, but as a scientist she can't publicly make this strong a claim, at least for now.

So, how are the little berries grown? How can we best decide whether to buy organic, conventionally harvested, or sustainably grown berries? How can we balance our quest for a low-sugar diet while enjoying the cranberry's inherent health benefits? And what can we do to ensure a healthy supply in the future?

The cranberry is a fruit which grows best on swamplands which can be overflowed at will with fresh water. It is an amphibious berry, which dwindles and becomes diseased if deprived of an occasional soaking. It is a God-send therefore to a people living in the midst of fresh-water ponds, and a third of whose land lay in worthless swamp, dear at a dollar an acre, useless to all, and owned only because it was part of the place.

—Charles Nordhoff, "Mehetabel Roger's Cranberry Swamp," 1868

2

THE FIRST CULTIVATORS

And ef it warn't for them cranberries you'd hev to go away this winter, little as you thought it, instead of sittin' comfortable in your own house. Tell ye what, boy, cranberry swamp's better'n goin to the Banks.

—Charles Nordhoff, "Mehetabel Roger's
Cranberry Swamp," 1868

For the first hundred and fifty years after the Pilgrims landed in North America, settlers treated the berry as just another native fruit to be gathered for a sauce or pie, occasionally to be used to make a poultice, or to be picked from local fields and carried on board ship. It was pretty much a laissez-faire arrangement.

Noting a need to establish order and, always mindful of the need to foster Christian devotion, English settlers in Provincetown, at the tip of Cape Cod, enacted an ordinance in 1773:

Voted, "That any purson should be found getting cranberries before ye twentyth of September excedeing one quart shall be liable to pay one doler and have the berys taken away."

Voted, "That they who shall find any purson so gathering shall have them and the doler."

Voted, "That any purson should be found getting cranberries on the Sabboth shall be liable to duble punishment."

Wampanoags, Aquinnah, and Mashpees on the East Coast had previously established their own rules forbidding the picking of unripe berries before the day assigned for harvest. One native, Samuel Sonnett, believed to be from the Pokanoket tribe, noted that the new settlers were dividing up the land and assigning portions to different members of their "tribe." He decided to act, and in 1702 he applied for and was granted ownership of two hundred acres of cranberry meadows and uplands on the south shore of Sampson's Pond in Carver, Massachusetts, an area where cranberries are cultivated today. Thanks to the Indian Land Grant, Sonnett and his family were allowed to fish, hunt, harvest cedar, and tap pine trees for tar and turpentine, all on lands reserved for the common use of land owners. To my knowledge, the Sonnett family were the only Native Americans with similar recorded rights at that time.

It took more than a hundred years for cranberry cultivation to begin. In 1812 in Dennis, Massachusetts, Henry Hall, a former schooner captain, returned from the sea to his farm nestled in the Cape Cod sand dunes. Winter storms, aptly named Nor'easters for the prevalent direction of their wind, regularly blew a layer of sand from the dunes beside Cape Cod Bay over the low-lying native cranberry vines on his property. Captain Hall began to notice that instead of being smothered, the vines sent shoots up through the sand. He decided to see what happened if he transplanted the vines to an area that was wet in the winter and dry in the summer, similar to the growing conditions for the wild berries. First he fenced off the area so that his cattle wouldn't trample the shoots. When vines began sending up new shoots that thrived and produced a group of small berries, he was probably as surprised and pleased as any new entrepreneur who experiences a "Eureka" moment that, at least in Hall's case, truly did bear fruit. Hall's neighbors joked about his wasting so much time building perfect conditions for a berry that grew naturally without any coddling from him or anyone else. Undaunted, he persisted. Four years later he was able to ship thirty barrels of fresh cranberries to the New York market. Then some of his neighbors took notice.

Soon, nearby farmers began to dabble in growing cranberries to supplement the income produced from raising cows and chickens and growing vegetables. If they couldn't afford to build their own family bogs, or didn't own land where the berry would flourish, they formed joint-venture partnerships with relatives or friends where they would contribute what they could afford, either in labor, land, or funds.

Charles Nordhoff, writing in 1868, described the process:

Enoch Doane read about the cranberry swamps in his agricultural paper, saw that the berries were in good demand in the Boston market, made a careful calculation overnight, and next morning rode out and bought a dozen acres of the worst-looking swampland in the neighborhood of Harwich. It took him a year to prepare a ten-acre lot. He had to cut drains, to build proper flood-gates, to clear the land of the rank growth of scrub oak which covered it, to cart away a foot deep of the sour top earth, to carry on new soil, to cover that with a layer of white beach sand, and lastly to set out his berries.

Growers began to be aware of subtle differences between wild berries picked from different locales, and they began to choose cutting vines with characteristics that would command a better price in the marketplace. In 1843 Eli Howes, another Cape Cod grower, transplanted a cranberry vine that produced hardy berries, matured in October, and had a firm, waxy skin that helped protect it from frost and damage. The Howes berry, named after its original cultivator, is the heirloom variety most frequently used as breed stock in today's high-yield hybrids.

Two other former sea captains, Alvin Cahoon and Cyrus Cahoon, began clearing land and seriously studying the best methods for cranberry cultivation. Alvin and his sons and neighbors labored through the winter of 1853 to dig, by hand, a five-foot ditch connecting two ponds in the middle of Cape Cod to lower the level of the water in one pond and allow for drainage to their newly built cranberry bogs. Today, Cyrus Cahoon is best known for developing the Early Black variety, so called because his wife is reported to have looked at the deeper shade of the season's first berry and commented, "'tis early, and 'tis black."

The success of the early cultivators began to be reported in local newspapers, and the word spread. Residents of New Jersey's Pine Barrens noted that the growing conditions in their area were similar to those on Cape Cod, and they had the same native berry growing wild in their sandy soil. They began by cultivating the smaller, local berry.

> Cranberries followed blueberries in the cycle of the pines. Cranberries grow wild along the streams and are white in the summer and red in the fall. In the eighteen-sixties and eighteen-seventies, people began to transplant them to the cleared and excavated bogs where ore raisers had removed bog iron, and that was the beginning of commercial cranberry growing in the Pine Barrens.

Around 1835 what is believed to be the earliest New Jersey bog was planted by Benjamin Thomas in Pemberton, an area that, like Carver, Massachusetts, was rich in iron ore and home to foundries producing farm tools and cooking utensils. By 1870 New Jersey produced more berries than Massachusetts, but the additional attention paid by the Massachusetts growers to leveling, weeding, and sanding their bogs paid off, and Massachusetts again produced higher yields. New Jersey growers paid attention. After devoting more effort to the leveling of their new bogs, they imported Howes and Early Blacks from Massachusetts growers, then planted the bogs with the higher-yield varieties.

Cranberry farming became "cranberry fever." In 1850 only twenty-six acres of cranberry bogs were cultivated in Falmouth, on Cape Cod. The wooden hand scoop, developed in 1880, enabled growers to gather the berries while moving across a bog on their knees, and by 1895 a full 3,255 acres of land on Cape Cod were devoted to cranberry cultivation. That was also the first year that Massachusetts's Plymouth County, with 3,766 acres of cranberry bogs, overtook the Cape in the number of acres devoted to growing cranberries. Between 1890 and 1912 the little town of Carver, Massachusetts, tripled the amount of land used to grow cranberries, and after 1905, the peak year for cranberry cultivation on Cape Cod, the yields from the Cape began to gradually decline. By 1945 the cranberry was the chief export crop from Massachusetts, but it was no longer dependent on Cape Cod bogs, which today are almost invisible except for some small, highly contentious acreage in Falmouth, Massachusetts that is still cultivated.

Charles Dexter McFarlin, another cultivator worth mentioning, is known less for his work in Massachusetts than for the part he played in developing the cultivation of cranberries in Oregon. McFarlin's father had regularly collected cuttings from wild berries. From the father's earlier research, Charles's brother Thomas successfully developed the McFarlin variety, a berry that flourished on his family's bogs. In a fit of wanderlust, Charles had traveled to

Oregon during the gold rush days. When a spring frost wiped out his Massachusetts crop, he headed back to the Oregon coast. There he noted that the growing conditions in Coos Bay, at latitude forty-one degrees, weren't too different from the growing conditions in Carver, Massachusetts, at latitude forty-two degrees. The main difference was a longer growing season. He sent a letter to his brother asking him to send some McFarlin cuttings west. After some experimentation with planting methods, the variety took, and is still harvested today in Oregon, Washington, and Wisconsin.

That cultivation of cranberries began in earnest on Cape Cod was no accident. The combination of iron-rich peat swamps adjacent to sandy hills or beach dunes provided the perfect growing medium for the native berry. Following the Civil War, the wooden schooners and packet boats that had plied the coast were replaced by steam-powered steel ships and railroads. Many of their captains and crew had been Cape Codders. They could no longer rely on fishing for their livelihood, as it was undergoing its own transformation from sail to steam, and hook to seining, both requiring fewer men. The seamen turned to growing cranberries, giving the cod a partial reprieve.

As true then as it is today, cranberries offer a higher return on investment than most other crops. Many a seaman owned land where the native berry grew so it wasn't difficult to transition to cranberry growing. New York afforded the best market for the berries, and dealers in Boston shipped their berries south to New York and Philadelphia for the highest prices. Benjamin Eastwood, an author and minister who wrote copious letters to the *New York Tribune* under the pen name "Septimus," wrote,

> The consumption of the cranberry in the great cities is such that the dealers can realize their own prices, by doing so as they did last fall, buy up the berry and get it into their own hands. The

wealthy will have the cranberry, and it is immaterial to them whether they pay eight or twenty dollars per barrel.

American cranberries, specifically from Cape Cod, had been sought after in Europe and Russia, where people's palates had grown fond of the sweetened yet tart flavor of the smaller *Oxycoccus* variety growing in their native soils. The larger American berry was even more desirable. The earliest shipping records for cranberries from New England to Europe date from 1550. Berries to be shipped were placed in small bottles filled with water, then hermetically sealed to maintain freshness. Eastwood remarked in 1856,

> We have seen a pint of cranberries marked "Cape Cod Bell Cranberry" sold at four shillings sterling, in Strand, London.

In 1832 the *Naturalist*, a publication circulated to the farming community, described Captain Hall as "taking the plants from their natural situation in autumn, with balls of earth around their roots, and setting them three or more feet distant from each other. In the course of a few years they spread out and cover the whole surface of the ground, requiring no other care thereafter, except keeping the grounds so well drained as to prevent water from standing over the vines."

Picture a crew member or schooner captain with this clipping in the pocket of his breeches, counting the days until he can be back on land experimenting with growing cranberries on his own Cape Cod acreage. He might have already made arrangements to purchase cut vines from Cyrus Cahoon, Eli Howes, or another grower who had already done the research and produced a healthy crop of berries. At home, the seaman will follow his neighbors' directions for sticking the cuttings into sand on bogs he had previously carved out of swampland when he was between sailings. And he will hope for the best. With the addition of some improved machinery, both Captain Hall's method and that of the hypothetical seaman prevail today.

It was the sage of Cedar Grove who said there was no satisfaction in anything unless you worked hard for it. Wood that warmed you in the cutting of it made a hotter fire in the airtight stove than slabs you bought at the mill. No rake invented for gathering cranberries ever picked them so perfectly as your own hands. "Look at those Hands of mine," he would say, "all scratched and skinned. But look, too, at the cranberries they picked."

—Cornelius Weygandt, *Down Jersey*, 1940

3

PLANTING

*The cranberry needs moisture, and that in great proportions for
such a small plant. If it is set out on the upland, and it does not
derive sufficient moisture from the atmosphere, that is, more than
is needed by surrounding vegetation, and if it cannot derive it
from another source, it will die.*

—Benjamin Eastwood, *Cranberry and
Its Culture*, 1847

USING PODS

It's thirty-one degrees Fahrenheit and overcast by the time I reach Mann
bogs, located between Route 495 and Head of the Bay Road in Buzzards
Bay, Massachusetts, on the opposite side of the Cape Cod Canal from the
spot where Henry Hall first began to cultivate cranberries. Keith Mann
and his father, David, grow cranberries on about 125 acres of bogs on
500 acres of land. The majority of their bogs are located down the hill
from their two houses. In nearby Wareham, Massachusetts, they grow
and harvest organically grown berries on 22 acres of bogs and uplands.
Adjacent acreage not owned by the Mann family exemplifies what this
area would be if the cranberry bogs weren't here. Small house lots with
single-family, mostly one-story homes and the occasional trailer extend

to the Mann property line. Within the Mann family acreage exists an oasis of woodlands, evergreens, uplands, and cranberry bogs.

Keith Mann meets me outside the entrance to three plastic-covered hoop frames, approximately 34 feet in width by 190 feet in length. He pulls back the wood-framed PVC door to the farthest greenhouse. As we step inside, I see rows and rows of ninety-six-cell flats, ten inches wide by twenty and a half inches long, holding what looks like three- to four-inch-high seedlings and stretching the length of the greenhouse. On this particular day, unusually high winds, clocked at sixty miles an hour, periodically shake the plastic, causing it to let out a rasping sound.

Keith started with about five hundred trays of Mullica Queen, a new hybrid that he purchased in bags as "stollons," cut vine sections, grown at Integrity Propagation's cold-storage room in Chatsworth, New Jersey. Each three-inch piece had been divided about four or five times, and each of those had about fifteen inches of growth. "So we just sheared them down and then cut all the pieces in the segments and replanted them to repropagate them." Keith says, allowing that it was a huge job. "We had a crew of about twenty people for about two weeks." The vines will be nurtured in the greenhouse until the middle of June or when the plants will have grown about six inches. "The cells will be completely full of roots by then," Keith adds, "so that when you pull the plant out, the whole cell comes out of the soil."

Mullica Queen's parent is the heirloom berry Howes. In the nineteenth century the movement of Howes berries was from Massachusetts to New Jersey, and at that time Howes was considered a higher-yielding berry than the New Jersey native. In the twenty-first century a descendant of the same Howes parent is returning to Massachusetts from New Jersey, where it is being bred to produce even higher yields. Eli Howes's berry is still known for its hardiness, and is thus a good berry for fresh fruit, where blemishes and indentations are not tolerated. These particular stollons will be grown organically and their berries sold as fresh fruit.

"I wish all my bogs were planted with some form of Howes," Keith says.

I ask Keith why he is planting the hybrid instead of the heirloom. "Greater yield."

Mullica Queen is a variety still in the research stage. For Keith to choose to plant it is a bit of a gamble. The hybrid is only four years old at this time, and the only substantial place where it had been planted when we met was on a half-acre bog on Integrity Propagation's New Jersey property. There it produced 400 barrels per acre. The Massachusetts state average yield was 178 barrels per acre in 2008, and that was the highest-yielding year in the history of commercial cranberry growing. The previous year's state average was 100 barrels per acre.

"For a four year old bog, that's exciting!" Keith states, referring to the 400 barrel yield. "I suspect it's going to get much better next year as the plants mature." Abbot Lee, the propagator, says that he'll be disappointed if Keith doesn't get 500 barrels per acre next year, and more in the following years. A five-by-five-foot test plot of Mullica Queens at the Massachusetts Cranberry Experiment Station in Wareham, Massachusetts, yielded the equivalent of 900 barrels per acre and allowed Keith to see how the berry functioned in a more northerly climate than New Jersey and in conditions comparable to his.

The upfront investment in a new bog is significant (around $45,000). Until a new variety has three to four years of time to grow in a new environment, no one knows how it will fare against local insects or in a particular growing season. "I'm guessing it will be five years before the plants are really established and we get a marketable crop," Keith estimates.

Marjorie Mann, Keith's mother, points out that most bogs are handed down from one generation to the next. "It's pretty expensive to try to do it on your own. I like to think our son is a fourth-generation grower," she adds. Marjorie's family had bogs in Norton, Carver, Falmouth, and Wareham. If cranberry growers had a First Family, Marjorie would be a member. Her grandfather was Irving Hammond. Her uncle, Robert Hammond, one of "the patriarchs" of cranberry growers, was a commercial grower at about the same time as Marcus Urann, the first president and cofounder of Ocean Spray, the cranberry cooperative.

Keith drives me to a one-year-old bog around the corner from the greenhouses. Tile drainage pipes stick out of the bog at intervals and drip water into the surrounding ditches. Use of buried tiles or "subsoil irrigation" more evenly distributes water throughout the bog, regulates runoff, and helps conserve water. "You're looking at about five miles of drainage tiles here," Keith says. He looks critically at the bog. "Look at that, they're all crooked. And I wanted a nice straight line that met an engineer's string at the ends."

This bog is another ongoing experiment. Its color is not the same dusty rust color of fully mature bogs at this time of the year. Keith plans to use organic growing methods on both this bog and the new bog, but he admits to problems inherent in growing organically. "Fertilizers, weed control are going to be real concerns, but we're going to give it a shot. Last year," he says, "we tried a biodegradable, organic mulch that we put down, and I was very excited about it." The mulch was promoted to accelerate plant growth and completely block out weeds. "This was going to be perfect!" He had planned to use the mulch on a small experimental area and the next year to use it on the entire bog. "Luckily, I did a very small experiment," he says. "In retrospect, it was a disaster!"

Keith describes his father scolding, "Why are you doing it differently? Everything was going perfectly."

"Just trying to learn," Keith says. A product of the Cornell School of Agriculture, he is young and not afraid to fail, trying to modernize yet not jeopardize production or the livelihood of his family and his workers' families. "When I was five years old, driving a tractor with my father, he asked me if I wanted to own the farm someday. Then after studying engineering in college, I came home to an industry that had crashed. I worked on the bogs without pay, and something changed. Now, I just want to be the best grower."

As we drive to the other side of a forested area on the property, Keith talks about creating a wind farm as part of his modernization plans. To our left is a pump station with a solar panel on the roof. Ahead is a mountain of fresh sand recently dug up to create the new bog where the seedlings in the greenhouse will be planted. At the top of the hill, we can look down to the left to three heavy loaders and other yellow digging equipment similar to the type used in the building of roads and highways. They're part of the equipment being used to renovate an existing bog, what will be the new home for the Mullica Queens. Beyond where the heavy equipment is parked, a bog is being renovated. The goal is to combine two irregularly shaped bogs, one that was previously a 1-acre bog and one that was a 2.2-acre bog, and square them up to make one larger, more efficient bog that will ultimately become an 8-acre bog. Below and to the right, bogs stretch to where they meet the woods.

To create the new bog, Keith is purchasing enough silt or clay to put a foot-deep layer over the smoothed-out sand. Then he will put eight inches of topsoil that has been previously stripped off the bogs and rests in piles to the side. Next, a six-to-eight-foot layer of sand will be smoothed over the topsoil. Lastly, drainage tile will be folded in every fifteen feet, as we previously saw on the year-old bog. Keith estimates three more weeks of heavy labor. To help defray the cost of the clay or silt, he has been trading sand from his uplands with a cement company. Clay is a byproduct from the process of washing sand, but it's not an even swap and Keith still has to pay upfront costs for the purchased clay.

From the days of Henry Hall to now, building a new cranberry bog has always been an experiment, hopefully a successful one, but not always. Drainage, how much or how little, how deep the ditches, how shallow the sand, remain to be answered anew by each cranberry grower each time he or she creates a new bog. Each bog builder wants to come up with the perfect formula. Keith is no different. He had previously taken the advice of an engineer who recommended that eight inches of loam and peat would hold the water. It didn't. "I was so frustrated because I did what the engineer said and it didn't work. And we spent a fortune." Keith stuck with the engineer but asked him for assurance that the present plan would work. The answer was to drive a lot of pipes into the bog, tamp the soil around them, then fill the pipes with water and measure the water drawn out of the pipes. The day after the pipes were installed, they were all empty. Keith laughs, as he relates the story, "Well then, it doesn't hold water," my engineer said.

After that, Keith and the engineer tried a lot of different soil depths. They finally agreed on a depth of twelve inches. That holds water. At a depth of a foot, water sits on the top of the bog and maintains the optimum degree of moistness to foster growth of the roots and vines. Keith added the drainage pipes to allow excess water to run off so that the roots and vines won't rot from too much water. It's a tricky balance, but he seems optimistic that this time he has it right.

Ideally, the top of a bog is one foot below the level of the surrounding land. Partly, this is to allow for sanding machinery and other equipment to roll down onto the bog. Keith's bog was sunk exactly one foot below the surrounding land, but he still wanted to bury drainage tile in the bog to make sure the roots weren't saturated with water. For environmental reasons, he also didn't want water running off the bog beyond the ditches. Most growers locate the tile drainage pipes one foot from the top of the bog, something Keith learned after it was too late to make the change. "And I had wanted to pull a four-inch tile in there," Keith adds.

On the new bog, he plans to place eight inches of soil and eight inches of sand over the clay, putting the bottom of the tile at sixteen inches below the surface. He'll also run the tiles lengthwise, which will be a lot less work than running them crosswise as was done on the year-old bog. "They pull in very fast and easy. You pull them in with a bulldozer fitted with a laser and plow." The plow digs the trench, the bulldozer drags the tile to fill the groove, and a laser ensures that the tile is level so that it drains properly. "What's been amazing with this bog is that usually you get some surface erosion that washes off into the ditches, and with this drainage tile, no water is sitting on the surface to form puddles and then erode the bog edges as it flows into the ditches.

"I have to admit liking innovating more than cranberry grow-ing," Keith says, pointing to a stretch of what appears to be a se-ries of tetrahedrons held by lengths of metal. "See those rusting booms over there. That's a project I started years ago. I hope to get back to it now that we have a little money to spend. It's going to be a 220-foot-long sanding boom, and it can be cantilevered from an excavator with counterweights and everything." If all goes as planned, the sanding boom will be capable of moving seven tons of sand at a time and sprinkling it over a bog, thus eliminating the need for ice to freeze so that sanding buggies can be driven over the iced-over bog. Keith has been collaborating with some engineering students and their professor at UMass Dartmouth. Every week he meets with three of the students who are using his experiment as their senior project. "It's really exciting," he says, his words coming faster. "We do computer modeling. It's really fun." The enjoyment is evident in his voice.

In July I return to find Keith and two biologists looking for worms they believe colonize in the piles of cranberry leaf waste left after harvest. We take his pickup and follow a truck holding flats of the Mullica Queens, now with six to eight inches of growth and the lanky, ranging look of teenagers waiting to be on their own. Honeybees are all over the mature vines we pass en route.

Today is one of the first days of prolonged sunshine in over a month of rain and gray skies. The temperature has risen to seventy-seven degrees. Keith stops to pick a few vines to show the men the many "pin heads and hooks" on the plants. He points out that five blossoms on a vine will usually produce one berry and that only half the uprights will normally flower in a year's crop. "Ideally, you want to see three-quarters to one inch of growth at the top. Like this." Keith runs his thumb and forefinger over the top portion of the new leaves and vine he holds in his hand. If the tips of the leaves are yellow, the plant has too little nitrogen; if the leaves are a mustard color, too little phosphorous. The leaves on the vine in his hand are varying shades of green.

We catch up with Tim Constantine, now unloading flats of Mullica Queen from the truck to a small buggy. When the cart is full, a man introduced as Run jumps into the driver's seat. Run, the tonal opposite of Ken, farmed oranges on his property in Cambodia before coming to this country. He is about half Ken's height and sports a thick head of curly hair to Ken's straight, wispy strands of blond. He moves with a seemingly effortless motion and is always just ahead of where you expect him to be.

Run drives the buggy onto the bog and heads out toward a low-slung construction consisting of two rows of two seats approximately four inches off the ground, each descending back from a tractor, an attached shovel at the front. A studded cylinder extends at ground level, behind the tractor and ahead of the planting operation being towed. Green vines protrude from sloped shelves in front of each of the seats, behind the driver, and from flats placed in the shovel.

Legs stretched out parallel to the ground, four men Run's size, all from Cambodia, heads covered against the sun's rays, sit in the four seats, facing the individual flats. A cotton canopy wide enough to cover the plants and the men droops over the edge of its frame to protect both men and plants from too much sun exposure. As I look out over the bog, the wagon perched in the middle of a four-inch layer of new sand reminds me of some kind of plant bazaar in a desert.

The tractor starts up and the whole contraption resumes moving in a slow, well-choreographed dance. The forward motion turns the cylinder, causing the studs to carve three-to-four-inch holes in the sand, one foot apart. Each man picks a plug of vine, soil, and roots from a tray. Then he reaches down to his left side and drops the plug into a newly formed hole just as his right hand is reaching for the next plug.

Richard Turgen, foreman for the Mann bogs, is driving the tractor and setting the pace for the planting operation. Richard is the third generation in his family to work on these bogs. His grandfather, Al Turgen, was the man responsible for developing the prototype of much of the machinery used by cranberry growers. David Mann credits Al with being responsible for the success of his family's bogs. "He could make anything and is probably the inspiration for my love of engineering," Keith adds.

On the other side of the bog, Ken is walking the previous line of plantings. He reaches down to gently push sand over any uncovered black soil or roots. "You want sand to be a reflector. If the sun reaches the roots, it will fry the plant," he explains. Pink flags dot the sand on the bog, marking the locations of 175 pop-up irrigation heads that were put in one and a half weeks ago. The temperature dropped close enough to the twenty-nine-and-a-half degree frost level two nights ago, July Fourth, to cause Keith to turn on the system. "Water is becoming a major environmental concern," he explains. To help conserve water, he normally irrigates at two in the morning instead of when the sun is up. In his truck a computer next to the driver's seat allows him to monitor the temperature throughout the farm. Within a year he hopes to install computerized probes that will also monitor the humidity in the ground at the root level of the plants. "I'm sure with the computerized irrigation system we'll use only half as much water." Temperatures can go down to eighteen degrees in the winter without harming the plant, but once the plant comes out of dormancy, the new growth is too delicate to sustain temperatures at that level. Then a grower will turn on sprinklers to warm the plants and prohibit freezing. Keith recalls occasionally seeing frost on the bogs in August, but not that often, and not in recent years.

In Massachusetts a cranberry grower is allowed to withdraw up to one hundred thousand gallons of groundwater and surface water per day, with a maximum of nine million gallons in a three-month period. Most growers use far less because they recycle water. In Oregon water usage is allotted only to growers who have previously qualified. An added incentive to Keith Mann's squaring off and leveling the Mullica Queens bog is that the Commonwealth of Massachusetts gives credits for bog leveling to conserve water.

USING CUTTINGS

The roots of the vines should be brought into close proximity with the muck below, that they may be stimulated to grow more rapidly. Women may be allowed to drop vines in this way, but they should never be chosen for pressing them into the ground. As a leaning posture is required, their skirts have a tendency to drag the vines out of place and waste them.

May 21 is sun-filled, with little wind—a perfect day for planting. From Route 44, about twenty miles north of the Mann bogs, I spot Barry Paquin, his sister, her boyfriend, and Barry's cousin on a bog I had previously noticed when a new layer of surface sand was being smoothed and graded. This is one of 150 acres of bogs the Paquin family farm in two states. As I walk to the bog's edge, Barry calls out, "Whoever can come along to help, we pay them ten dollars an hour."

Unlike the process on the Mann bogs, utilizing cells with seedlings and roots, Barry and his helpers are replanting an older bog with mowed vines. A grower in nearby Hanson sold Barry the pruned clippings from one of his existing bogs. Clumps of vines the size of sage balls fill the back of Barry's pickup truck. Each vine is approximately six inches long, some divided into two branches. They are Ben Lears, a variety grown primarily for juice. Barry chose to plant the Ben Lears on this particular bog because the variety does well in damp settings similar to the conditions on this low-lying peat bog.

"We got these vines from a good producing cranberry bog. That's why we paid two thousand dollars per ton. We bought eight tons to plant four acres," Barry tells me. Peter Paquin, Barry's father, bought his first bog in 1978 and started planting his own bog in 1980. When Route 44 was built, it bisected the two bogs, disrupting the natural flow of water, and drowning the vines in the summer months when they needed to be exposed to the sun. Parts of the bog on the north side of Route 44 were replanted with houses in 1990. "Everybody likes to plant Stevens, but Stevens like to be dry. Ben Lears like to be wet," Barry adds. Stevens is a higher-yield variety, older than Mullica Queen, but newer than Howes. Despite the eighty-four-degree temperature, Barry has managed to keep the vines moist by means of a sprinkler. He irrigated the bog prior to sowing the vines, he will irrigate the bog again this evening, and he will continue irrigating until the vines put down new roots.

Barry and his family team walk the new bog, separating the vines from the clumps, then sowing them evenly with the same easy swing and rhythm used to scatter birdseed. Yesterday, they completed the center of the bog. Now they are working the remaining portion of the far edge. They began on Saturday and have worked all day each day since, with the exception of Sunday morning, when it rained. Today is Thursday, and they hope to finish the planting in a few hours.

Barry picks up a "vine setter," a three-foot-wide series of disks attached to a roller, a handle running the width of the roller, and a motor at one end. He guides the vine setter over the newly sown vines as the machine's parallel disks gently push the vine ends into the ground. Given the required amount of water, sun, and care, the vines will do the rest. If all goes well, they will produce a new crop of harvestable cranberries in three to four years

As I turn to leave, Barry starts the motor on the vine setter to plant the remaining vines. Meanwhile, in upstate New York, near the Canadian border, his father is sitting in the seat of a sixteen-foot-wide vine setter with windows and a rain drain as it moves across a fifteen-acre bog he is replanting. Mechanization has allowed one man to replace families of immigrants on their hands and knees placing vine cuttings in holes while the bog foreman followed wielding a "dibble," a multipronged instrument with a wooden handle designed to gently push the vines through the sand to the layer of peat below. In the spirit of using what there is to work with, "Peg Leg Webb," the oft-cited nineteenth-century New Jersey grower, is reported to have used his wooden leg as an oversized dibble when planting his vines.

After leaving Barry, I stop to see how Israel is choosing to replant the Beaton bogs. We hop into his pickup and drive down a rutted sandy road to an area of ten to twelve bogs with no buildings in sight. Israel stops at a bog the Beaton crew planted in June the previous year. This bog was planted with Stevens in a manner similar to Barry and Peter Paquin's method, that is, with cuttings instead of purchased pods, but the Beatons are taking no chances on the quality of the parent vine and planted this bog with prunings from another Beaton bog also planted with Stevens. In a sense, these vines are second-generation Beaton vines. The bog could produce small berries this year, but "I don't want to harvest them until the roots get stronger," Israel tells me.

He points to a patch of green in the middle of another bog. "Those vines, my Ben Lears, came out of dormancy in late April. They come out first. That patch over there, I saw bells on it this morning."

"Bells?"

"That's what I call 'em." He leaps onto the bog to pick me a sample vine. "See, if you look at the bells," he indicates the budding berry drooping from the vine, "that's what they look like." He stoops to gather a handful of raw berries and, after offering some to me, pops them into his mouth as we head back to the truck.

Early growers divided the different varieties of wild berries into three types based on the shape of the berry: the bell, bugle, and cherry. Of course, shape has traditionally been less important than color with regard to cultivated berries, and the darker color has commanded a higher price. The Early Black variety grown at Piney Woods has a uniformly dark red color, matures earlier than other varieties, and sports a round, cherry shape—all qualities promising a higher price at market.

EARLY RECIPE FOR CRANBERRY JUICE

Put a teacupful of cranberries into a cup of water, and mash them. In the meantime, boil two quarts and a pint of water with one large spoonful of oatmeal and a very large bit of lemon peel. Then add the cranberries and as much fine Lisbon sugar as shall make a smart.

—Elizabeth Wollsey, *Compleat Cook's Guide*, 1683

4

IRRIGATION

If you strung all the cranberries produced in North America last year, they would stretch from Boston to Los Angeles more than 565 times.

—P. Ann Pieroway, *Taste of Cranberries,* 2007

On a cranberry bog, spring is the most demanding of the seasons. The winter nap is over. The protective winter flood has been drained, and the vines have broken out of their dormant period. The plants are in a fragile stage, with delicate buds just beginning to grow and seek the sun. Up to now, when they have been slumbering, the cold weather hasn't been a danger. Soon it will be. Frost, the enemy of the cranberry, lurks ready to destroy previous months and years of work.

By definition, a bog is lower than the surrounding lands. Bog temperatures can be as much as fifteen degrees colder in the late evening and early morning than temperatures for upland areas. Weather forecasts are usually for land at sea level or higher, and frost tolerances vary for different cultivars or berries. Plants with large buds such as Ben Lears and Stevens can not tolerate temperatures below thirty degrees Fahrenheit once the terminal buds begin to elongate; smaller buds such as Howes and Early Blacks can tolerate temperatures as low as twenty-seven degrees at the same stage of development. When the fruit is fully red, it can tolerate a temperature as low as twenty degrees. To be safe, most growers turn on their sprinklers when the bog temperature dips to thirty-two degrees.

Throughout April and into May and June, growers in the Northeast and Wisconsin are on alert. One wrong decision or pump failure and the year's crop is destroyed. In early April, to protect the berries, growers put sprinkler heads on their bogs. Water sprayed from the sprinklers protects the vines from frost and irrigates the bogs during a dry spell.

Because of the difference in terrain, frost may have been less of a problem for the Cape Cod pioneers who first cultivated the berry. Few trees grow on sand dunes close to the sea. Most cranberry bogs today are inland, surrounded by uplands and evergreens that tend to keep the cooler air on the depressed level of the bog and increase the chance of frost on the plants.

Irrigation for frost protection is not limited to the spring period when the plant is most fragile. If there are plumes of water spraying over a cranberry bog on a cold day just before harvest, chances are that the water is protecting the berries from frost, although a ripe cranberry can withstand temperatures as low as twenty-three degrees without damage.

Nineteenth-century pioneers learned early on to protect the new growth from frost if they wanted to have any crop by fall. In 1840 a grower named Jarvis Lovell attempted to keep frost off his vines by suspending yards of cotton on wooden posts stuck into the bog. Salt marsh grass, readily available at the shore and packed around the plants was another remedy. What growers soon learned was that water was the most reliable frost protector. Whenever possible, they built dikes and dams with removable boards that could be lifted to allow water to flow over the bog to protect the vines. Today's bogs don't look that different. They are still surrounded and bisected by lines of water. The difference is that they are now also dotted with sprinklers, some basic, some hooked up to computerized monitoring systems.

Clusters of white petals with pink stamens and oval leaves dot the shad bushes at the edge of Iain Ward's nine acres of bogs in Lakeville, Massachusetts. On the day of my visit, a yellow ball of fluff tumbles out from beneath one of the bushes at the edge of the bog. Then it regains its balance and scurries off, only to be joined by six other wild turkey chicks and, lastly, the clucking mother.

Four houses surround Iain's bogs on two sides. The house he shares with his wife, Christine, and two sons is on the hill at the southern end; a wooded area belonging to his family abuts the western side. Three other homes and a highway define the boundaries on the remaining two sides. A grower with holdings the size of Iain's is considered a small grower. By virtue of his size and location, he lacks the usual buffer zone to shelter his operations from the concerns of his neighbors. In heavily populated areas such as southeastern Massachusetts, increasing population density is driving an increase in regulations. And a lot of growers like Iain aren't happy with the results. "We're getting regulated on muskrats. We're getting regulated on box turtles," he says. "We pay taxes, we provide employment, and we provide a park for our neighbors." Pointing to the reservoir across the street, he adds, "That water level is kept at a steady level by the dike that I maintain. It's full of fish and other wildlife, but it was created to provide water for my bogs. If I didn't maintain it, the dike would fail and the water would be gone. And the fish and other wildlife would be gone with it."

In the Northeast, small growers with under twenty acres of bogs outnumber midsize or large growers, although the largest amount of land is held by a few growers of two hundred acres or more. Most cranberry growers agree that the small grower has a difficult job. Iain doesn't pretend it is easy. His cost and risk is borne by fewer acres than that of the larger grower. A dry picker, dump truck, sanding buggy, bulldozer, and tools came with his purchase of the property, but he "hires out" picking at harvest time, and he sold off the dump truck and bulldozer to cover the expenses of the critical first year. Iain has neither the inclination, the storage space, nor the acreage to warrant owning big equipment. When he needs an excavator or another piece of heavy equipment, he rents it.

The Ward bogs, like many older Massachusetts cranberry bogs, are irregularly shaped. One bog is round. Working these bogs is a less efficient operation because modern equipment is designed for rectangles, where a grower can run his equipment up and down the bogs in straight lines.

A grower with acreage the size of Iain's needs a second job. For Iain that job is his environmental consulting business, New England Consulting Services (NECS). But he has to be available for both his clients' needs and those imposed by nature on his own bogs. It's a delicate balance. He has a degree in environmental science and previously worked at the local U.S. Department of Agriculture's field office, where he helped put together farm plans for cranberry growers. "Driving around with all the growers in their pickup trucks, I learned so much," Iain says. Today, growers hire him to negotiate the regulatory process; work with local, state, and federal officials; and assist with land-use planning.

On April 29, Iain, his four-year-old son Caleb, and I walk down the hill from their house to their pump house. While Iain is giving permission to Caleb to show me the correct way to install sprinkler heads and irrigate a cranberry bog, Caleb's three-year-old brother, Casey, and Bowdoin, the family dog, run down the hill ahead of us.

Caleb instructs me: first you unlock the door to the pump house, then you raise the shutters to allow ventilation for the Ford gas engine that operates the system. Caleb points to the needle in the center of a gauge attached to the engine to show me how he checks the oil level. He looks to his father for approval. Each bog has its own water valve. Once the pump is primed, we head to the bogs, where Iain and Caleb reach over the bank and down through the grass to find the valve for the first bog scheduled to receive sprinkler heads. Caleb, helped by his father, turns the valve clockwise to open it. Water bubbles up in spurts every ten to fifteen feet throughout the bog, showing Iain and Caleb where to attach the aluminum risers topped by brass sprinkler heads. Iain holds his hand over one of the fountains of water, then secures the head by twisting it into the existing cup with his other hand.

Traditional sprinkler heads were made of brass and have become a target for thieves. Pop-up plastic heads that retract into the ground are available and offer a possible solution, but the expense to renovate the irrigation system make it prohibitive for most small growers like Iain. After Iain has attached a complete row, enough pressure builds to turn the heads and send circles of spray around each sprinkler. Later, when the sprinklers are all in operation, the pump will keep the pressure at forty-five pounds per square inch to ensure a continuous spray of water.

As we walk off the bog, Iain picks a shoot from one of the vines to show Caleb and me the story he can read from it. The darker leaves closest to the top are from last year's growth. From the center of the top two leaves, a fledgling green bud is barely discernible. The prior year's growth is visible in the green leaves lower down on the vine. Noting two year's of growth still on the vine instead of just the previous year's growth, Iain worries that he may have provided too much fertilizer to the plants last year. "And that's the whole art of growing fruit," Iain explains. "The plant, being a vine, wants to trail along the ground. If I apply a layer of sand, anchoring the runners, the plant produces shoots that grow upright to bear blossoms and fruit.

Because it lives to reproduce its own species, it also produces fruit to seed. I want to trick it into thinking it's under stress [prohibiting it from growing along the ground] because that's when it produces a berry. Timing is critical. If I get the proper balance, the uprights bear fruit, and the vine stays healthy for next year's crop." Ideally, a grower wants to maintain a consistent yield from year to year. If the bog is sanded or fertilized too soon or too late, too much or too little, the yield might be high one year and low the next.

Iain and his family are a special entity. They chose this way of life. Iain's wife, Christine, is a descendant of Capt. Cyrus Cahoon, one of the families of ship captains who combined whaling and cranberry growing on Cape Cod. Captain Cahoon is credited with developing the Early Black, the heirloom variety growing at Piney Woods and here on Iain and Christine's bogs. Cranberries are in Christine's blood, and if she and Iain have anything to do with it, it will be in the blood of Caleb and Casey.

Bowdoin is running in circles. The boys are squealing with pleasure and soon soaked. Iain looks over at them, laughs, and shakes his head. "What can I say?" He admits that some people would ridicule him for allowing the boys and the dog to run back and forth on the vines, but "In my mind, it's better to have them out here if they're having fun." He points out that Caleb weighs only forty-eight pounds, his brother thirty-eight. "The damage that they're doing is insignificant in contrast to the long-term benefits to their growth."

We head back up the hill with two soaking wet little boys. "Around here," Iain says, "If you want to raise your children somewhat agriculturally, I can't think of anything better than what we're doing right here, right now." The cell phone rings and Iain sets up an appointment for his consulting business. This is what permits his way of life to flourish—that and his and his family's inventiveness and desire to make it work.

Just over the Carver line, Wareham Road cuts off to the left. The DO NOT ENTER sign belies the road's name, indicating this might be the main road to the next town, as it was for many before Route 495 split it in half. Thanks to the fortune of highway construction, the road and everything to either side of it is the private enclave of the Rinta family. The family owns and harvests thirty-two acres of bogs. Like the location of their road, they are betwixt and between: a little larger than most small growers, but still small enough to do most of their own work.

The weather is eighty-four degrees with a light but steady wind as I approach the first bog on the property. Bluebird boxes dot the perimeter, and warblers flit across the road to the surrounding woods. Andrew Rinta and his dog, Meda, appear from behind the pump house at the edge of the Weweantic River, the source of water for these bogs. Below the bank, a rushing stream cascades over rocks and on toward Buzzards Bay. Intermittently visible in the water's swirls is the top of the pump's suction box. Andrew brings his four-year-old son here fishing for the bass he says are plentiful. "I never was a good fisherman, but I can sit here on the river bank, with my son, just catching one fish after another."

Growing up, Andrew watched his father spend long hours for little money while harvesting cranberries. And he wanted no part of it. He left to go to art school, then served four years in the marines, got married, and opened a tattoo shop in a nearby town. His body is a moving billboard for his previous business. In 2004 his father called and mentioned a ten-acre parcel that lay next to the Rinta family bogs. The land was for sale. Andrew's father had been given the right of first refusal. "This was the land where my brother and I grew up. I hated to see it developed," Andrew says. "In a second, I said I'd take it. I hadn't even told my wife. I just said I'd buy it."

Now, he can't imagine any better life. He has two children, three and six; he built a home on a hill next to the bogs; and his wife teaches in the local schools. Unlike many other growers with similar acreage, this family has never had to sell off the uplands surrounding their bogs. They have no neighbors calling in the middle of the night complaining about the noise of generators being used for frost protection, no neighbors to complain about helicopter noise during harvest, and no neighbors bringing suit for fear of chemicals in the pest management.

"They're not making any more land," Andrew says wistfully, "and I'm fourth generation. There's a kind of mystique about it. If my children would decide to get into it, that would be cool. But I'm not going to push them any more than my father pushed me. I just hope I can preserve this way of life so that they have the option to choose." Andrew's great-grandfather Walter Heleen was part of the early waves of immigrants from Finland who worked on the bogs. He came to this country in 1899, when over 1,300 acres of cranberry bogs were under cultivation in Plymouth County, and the new industry required immigrant labor. After hiring himself out as a cranberry picker in Carver at harvest, he decided to stay. Later, he would meet and marry a local girl who bore him four sons and two daughters. The family—children, women, and the men—pitched in to work on the bogs and pooled their earnings to eventually purchase their own bogs and grow their own berries. Andrew didn't want to lose ties to that legacy.

The pop-up sprinkler heads at Andrew's bogs, together with his and his family's ownership of the surrounding land, provide a tremendous advantage. Unlike Iain Ward, who was up at three in the morning worried about frost for much of the month of May, Andrew can set the gauge in the pump house and rely on the pop-up sprinkler system to automatically protect the new buds when the temperature falls below freezing.

Andrew gently moves leaves and vines away to reveal one of the new plastic pop-up heads for my inspection. By producing a finer mist, pop-ups use less water than the old sprinkler heads. But to function the pop-ups require a new or upgraded pump system. Many growers are taking the opportunity to also install solar power as part of the renovation. Andrew sees that as part of next year's plan. For this season he has installed a new filtration system on his old pump. The cost to convert to pop-ups is $4,500 to $5,000 per acre. And that doesn't include the cost for a computerized system.

We walk to the pump house he built for his new irrigation system. Putting the pieces together has been a two-year project where Andrew, like most cranberry growers, has custom-made the machinery to fit his particular bogs' requirements. He purchased the filter, then plumbed it and figured out how to make it fit with the rest of the system. New valves allow him to isolate which bog he irrigates or to limit a pesticide application to only the infected area. Andrew has set up the new system so that it connects from his cell phone to the operating computer. "So I can be out to dinner with my wife," he explains, "and I can get a frost message. Then I can call my pump, check the bog temperature at the pump house, and start the pump through my cell phone. You never know what's going to go on in the middle of the night. A carburetor will let go on a frost night, and you've gotta know how to fix it in a rush." He doesn't mention that it helps to know how you built the equipment. The old system operates with a 1990 Chevrolet engine that runs on propane; Andrew built the new system around a new Chevy coil pack engine. "This is my latest and greatest," Andrew laughingly tells me as we open the door to his new pump house. "It's too bad you can't ride it. You can't drive it. All it's going to do is pump water for the rest of its life. And I was just like a little kid when it was delivered. I sat on it, I stood on it. I hugged it."

Before he could build the pump, Andrew had to build a home

for it, beginning with a cement mixer and the 103 bags of Quikrete needed to build the slab. He points to a spot on the wall where a computer panel will be placed to enable him to "talk" to the pump and allow the pump to "talk" to him. Tomorrow Andrew will hook up the pump, minus the control panel, allowing him to manually start the system.

As water is sucked up from the river, the new filtration system screens out any algae, dirt, and remnants of debris before it reaches the pop-ups. Andrew has built the new system so that it can handle twenty-five acres of bogs, ten more than he presently has planted. For now, he has all the bogs he wants to manage. The Rintas own eleven acres of nearby woods. That land acts as a buffer for traffic on Route 495 but has the potential to become additional cranberry bogs.

At the bog across the street, three men clean debris and old vines from the ditches. They and three others work for Andrew in season. Unlike some of the smaller growers, he always hires the same men. He hires himself and the men out to smaller growers who might not have the equipment that he has by virtue of greater acreage and a longer family history. "Frankie, one of those guys, he's been with me since I was fifteen," Andrew tells me. "I grew up with these guys. They know the bogs and they know what to do. It's not like retraining somebody every year."

In March a typical vine displays a tiny "cabbage head," similar to the one Iain Ward pointed out on his bog. That cabbage head is the bud for that year's growth. After harvest it's possible to walk onto a bog and see the cabbage head for the next year's crop. The vine keeps its leaves. When new leaves are green, the plant is breaking out of dormancy, a process growers refer to as "bud break."

By May 21 Andrew's bog is just beginning to green up. A red tint is still visible. Spots of green indicate an area that might benefit from more direct sun. Andrew admits to having "somewhat shot myself in the foot last year." He means that to put in the new irrigation system after harvest, he and his crew had to drive a truck over the vines. It's always a gamble to put in a new system and abandon the old ways, but Andrew is confident he has made the right decision. "We're going to be the new farmers," Andrew says. "I don't need to wear overalls and suspenders and be stuck on the property full-time. But my wife and my kids are the goal," Andrew tells me as we walk back toward my car. "That's where I'm really fortunate."

Agriculture is, by its very nature, brutally reductive, simplifying nature's incomprehensible complexity to something humanly manageable; it begins, after all, with the simple act of banishing all but a tiny handful of chosen species.

—Michael Pollan, *Botany of Desire*, 2006

5

A MATTER OF INSECTS

The hallmark of life is this: a struggle among an immense variety of organisms weighing next to nothing for a vanishingly small amount of energy.

—Edward O. Wilson, *The Diversity of Life*, 1992

Cranberries are an ornery bunch. They don't recognize a twelve-month calendar. They respond to the length of days, change of seasons, and months of the year, but they have a sixteen-month growing cycle, not a twelve-month cycle. The only way the plant can correspond to nature's cycle of months and seasons is for several stages of growth to take place at the same time. In July a terminal or new bud will begin to grow at the top of the plant at the same time as the flowers or "bells" are forming a hook and also blossoming halfway up the stem.

Throughout the summer and fall, new floral buds are developing in the terminal bud as fertilization, growth, pollination, berry development, and harvest are taking place on the same plant. In a sense, two sets of berries are growing on the same vine, but on different schedules. So how do growers balance the needs of this year's crop without damaging next year's?

First, they must decide whether to grow organically, sustainably, or conventionally. The former is the most labor-intensive, for many the purest and the most difficult to maintain. Sustainable growing respects the environment while diminishing some of the labor of organic farming and still producing a reasonable return. Conventional cranberry growing is the most efficient method and produces the highest yields but requires a delicate balance of herbicides, fungicides, and pesticides.

By definition, a bog is composed of layers of decaying sphagnum moss (peat). Because of its makeup, it lacks the nutrients required for most plants to flourish. Nitrogen is an essential component for cranberry growth. On a peat bog, nitrogen is in the air and soil, but requires either microbes or a fertilizer to release it to the plant. Growers who rely on only naturally occurring microbes are limiting both the size and yield of fruit. They may also be increasing the likelihood of rot to the vine. To augment the nitrogen too generously is to limit the hardiness of next year's growth. Timing and amounts are critical to berry production not only for the present year's crop but also for the following year's crop. That's where fertilizers enter the picture.

The second component, and the true definer of organic, sustainable, or conventional, is PESTICIDES, the word that has given cranberries a bad name. It's generally agreed that the first grower to use some combination of substances on or around his vines to deter insects was a James Lovell on Cape Cod. In 1844 Mr. Lovell is reported to have spread lime, salt, and wood ashes on his bogs in an attempt to control the cranberry worm. The results were mixed, but his fruit did win a cash prize at the Barnstable County Fair.

By the mid-1850s, after cranberry cultivation had moved beyond the Cape and was beginning to be seen as a serious means of livelihood, various "experts" were challenged to come up with a means to eliminate the cranberry worm or at least to block it from munching on cranberries. Professor Louis Agassiz, the esteemed namesake of Harvard's Peabody Museum, recommended that growers divert the fruit worm from the berries by building fires around the bogs so that in its adult stage the insect would fly into the fires. Fortunately, Professor Agassiz's reputation is not dependent on his advice to cranberry growers.

Three years later Augustus Leland noticed that when he flooded his bog into the summer he drowned the cranberry worm, then and now the number one insect plaguing cranberry growers in the Northeast. For the next fifty years growers relied on sanding and flooding to try to combat insects, weeds, and fruit rot. Then in the early part of the twentieth century, chemistry began to replace water and sand. By the late 1930s sodium cyanide, copper sulfate, sulfuric acid, pyrethrum, and lead arsenate were being purchased by the barrel, pumped through hoses, and dragged across the bogs by immigrant workers who then hand sprayed them onto the vines.

Wars often provide a testing ground for the use of new chemicals and the development of new equipment. This fact wasn't lost on the observant cranberry grower. At the end of World War II, growers began to hire airplane pilots to spray DDT, the potent chemical that had been so effective against insects believed to carry typhus. Ocean Spray's founder, Marcus Urann, upped the ante by introducing the first helicopter spraying of DDT.

Until Rachel Carson wrote *Silent Spring* in 1962, helicopters spraying DDT and various other powerful chemicals accounted for 50 percent of the insect control on cranberry bogs, and the process was virtually unregulated. *Silent Spring* alerted the public to the fact that DDT not only caused the deaths of osprey, peregrine falcon, and various songbirds but stayed in the soil and in the fatty tissues of birds, fish, and mammals where it was ingested by their predators. More than five hundred thousand copies of *Silent Spring* sold in its first five years. The public reacted, and in 1970, under the presidency of Richard Nixon, the Environmental Protection Agency was formed. The next year, Massachusetts banned DDT. The EPA followed by banning Chlordane, Silvex, and Dieldrin.

Helicopters still spray fungicides and pesticides on some of the larger cranberry bogs today, but most growers prefer to treat their bogs in a manner more respectful to the environment. Monika Schuler is a consultant in a field termed "integrated pest management" (IPM). She is also a small grower who knows firsthand what it takes to produce the round, uniformly colored berries that we take for granted when we see them on the supermarket shelf or at a farmers' market.

"If you are a grower and you have to apply fertilizer, and the bag says to apply so many bags of nitrogen per acre—and I have actually lugged those fifty-pound bags across a bog while hoping not to damage next year's growth—and if you try and do a couple of those bags per acre, it takes you hours and days. And the sweatier the better," she emphasizes. "After that you can spend more days working on your broken-down machinery that nobody else can fix but you because you built it originally."

The Decas company, a family of cranberry growers that once supplied herbicides, pesticides, fungicides, and fertilizers to other growers, was the first to offer commercial IPM services to the industry beginning in 1981. One of the early certified IPM managers hired by Decas had previously managed the grounds at the White House. After a few years, when Monika decided to return to Washington to begin a PhD program in marine biology, John Decas offered the job to her. The pay was six dollars an hour. By the end of the summer, as she describes it, "I was hooked on cranberries and the world of cranberry growers." Today, she walks about 1,100 acres of bogs each summer, sweeping the vines while looking for pests.

"I recognize how incredibly fortunate I am," she says. We are sitting in her kitchen with an unbroken view of her children and

their friends playing outside on fourteen acres of farmland in Mattapoisett, Massachusetts. During the growing season, Monika often
works eighteen- to twenty-hour days, but she acknowledges that the
tradeoffs far outweigh the negatives. "This way, I can be here with
my kids."

"I had a lot of very good teachers," she insists. "It's one thing to
read in a book and see what a bug is supposed to look like and what
kind of damage it can cause. But to actually spend time with the
growers, my clients, and to walk the bogs with them as they point out
what is going on here or there . . . That's the way to best understand
what is happening."

If Monika sees something at a bog that doesn't look right, she calls
Carolyn DeMoranville at the Experiment Station of the University
of Massachusetts. "Between the Experiment Station and folks in the
business, we usually figure out what's going on." Her clients are all
owners of Massachusetts bogs, the most distant being those managed
by Tom Larrabee on Nantucket.

Monika agrees with grower Gary Weston that there probably
is more stress to the vines from today's method of species-oriented
treatments, where each insect is treated separately. This technique
is considered to be less damaging to the environment than the older
method, where a lot of growers would spray a general pesticide by
May 30. It is also more expensive, for growers now pay the same
for treating each species as they paid for one overall treatment. "If
you have fifteen different bugs on your bog, and it costs you fifteen
dollars each to spray, that adds up," Monika points out. It's a small
investment to pay, however, to avoid a replay of November 9, 1959,
the cranberry growers' "Black Monday."

It was a week before Thanksgiving, and grocery stores were
stocked with cans of cranberry sauce in anticipation of the holiday,
that time of the year when 95 percent of the cranberry crop was con-

sumed. Up to that moment, cranberries had enjoyed an unsullied reputation as the purely American fruit with a Northeast accent, grown by the trusted farmer in plaid shirt, overalls, and rubber boots. Then Arthur S. Flemming, the U.S. secretary of Health, Education and Welfare, held a press conference, where he made the following announcement:

> The Food and Drug Administration today urged that no further sales be made of cranberries and cranberry products produced in Washington and Oregon in 1958 and 1959 because of their possible contamination by a chemical weed killer, aminotriazole, which causes cancer in the thyroids of rats when it is contained in their diet, until the cranberry industry has submitted a workable plan to separate the contaminated berries from those that are not contaminated.

Asked by a reporter if that meant that cranberries in other states were safe to eat, Secretary Flemming answered that his department had no way of knowing in what parts of the country the tainted berries might appear. Overnight, stores around the country removed all cranberries from their shelves. Orders that had been placed and awaiting delivery were canceled. Flemming's timing couldn't have been worse for the cranberry growers. The chemical genie had been let out of the bottle. Thanksgiving of 1959 was a holiday without the traditional cranberry sauce served alongside the turkey.

Two years earlier, in the December 1957 Ocean Spray newsletter to its growers, Kenneth G. Garside, then general manager of the cooperative and a prescient environmentalist, had written an article titled "Why the Excitement over Aminotriazole?" In the article he warned his fellow growers that disregarding the inherent dangers of the weed killer could "cause and result in the condemnation of your crop."

By the time of the scare, the Department of Agriculture had approved use of aminotriazole as an herbicide with the stipulation that it be used on bogs only after the harvest. It had been applied by the growers of the suspect berries before the berries were picked. To this day, many growers in the East feel that they were unjustly damaged and that no berries from New England were sprayed with the herbicide. Three days after the press release, a total of eighty-four thousand pounds or eight lots of cranberries were found to have a coating of aminotriazole. One of the lots was from a Massachusetts bog, four from Wisconsin bogs. The analysis was paid for by cranberry growers. President Dwight Eisenhower announced that growers who could prove their berries free of the herbicide would be compensated for their crop. A total of $8,000,500 was paid in reparations, but the damage had been done. That year's crop was destroyed, and many family bogs were lost or sold at bargain prices.

Unlike yesterday's pesticides, those used today are the result of extensive research geared to understanding how and when a particular insect lives, breathes, eats, and mates. After the research phase, a chemist might produce a pheromone or perfume that masks the scent of the bug's potential mating partner. When the male has flown from object to object in search of a receptive female, only to be thrown off by an offensive scent, he is too exhausted to mate and produce berry-eating progeny.

One of the dangers Monika sees is in the hybrid, high-yield vines or pods developed in one climate and shipped to a different climate. Only after those vines begin to grow will the farmer who bought them, built a bog for them, and is now growing them realize that those plants or vines may have brought with them the eggs of a new and dangerous invasive species, one from a different environment where the period of cranberry budding may not coincide with budding in its new home, and thus the species may not have been a known enemy of the cranberry.

As insects adapt to a warming climate, they are moving toward the poles at a rate of more than 1.6 miles per year, roughly paralleling the rate of climate change. The rate of movement for molds and other fungi is even more rapid (4.16 miles per year), thus changing their phenology, or the stages of their development, relative to that of the cranberry. It is generally believed that loss of crops to various pests accounts for between 10 to 16 percent of each year's total crops worldwide. Given concerns about feeding the world's population, any change that could put more pressure on edible plants is a serious concern.

Each state issues regularly updated guides and regulations for use of fungicides and pesticides. Monika's job includes recommending only those methods of treatment approved by the state and federal government. Occasionally, she has to inform a paying client that the grower must pay to clean up a pest problem in a manner that conforms to existing regulations.

Despite her love of being outdoors on the bogs, Monika's job is not without anxiety. Her trained eye and her experience can put money in a grower's pocket come harvest time or, if she misses a potential infestation, can cause the grower to lose all or part of a crop and a year's income. "I've been on both ends of the argument," she acknowledges. "I know what it costs to grow cranberries, but I also know what it's like to not have enough money to keep a house running or children fed."

Even until the breakout of the Eastern War, there were to be seen among the bales of hides, hogsheads of tallow, bundles of bristles, and bales of hemp, certain quaint-looking earthen jars, which contained cranberries for the use of the lords and ladies of London. . . . Now that the Crimean War has effectually put a stop to the importation of Russian cranberries, it is but reasonable to suppose that the American article will monopolize the English market.

—Benjamin Eastwood, *Cranberry and Its Culture*, 1847

6

ORGANIC, SUSTAINABLE, OR ALMOST CONVENTIONAL

As Mike Heath showed me around his farm, I began to understand that organic farming was a lot more complicated than simply substituting good inputs for bad. A whole different metaphor seemed to be involved.

—Michael Pollan, *Botany of Desire*, 2006

ORGANIC

Another rainy day. Fifty-nine degrees with drizzle, and I am beside Route 58 in Carver, Massachusetts, where Dom Fernandes raises the Early Black variety of cranberries with the Massachusetts Certified Organic stamp approval. Today is June 24. The bog is green, spring green, vines loaded with buds just at the point of flowering. Dom, by Massachusetts standards, would be considered a medium-size grower. He describes going to a meeting in Wisconsin, where he was asked how many acres he farmed. "When I told them about twenty-five, they asked me what else I did for work. I told them this was all I do, and they just looked at me with this blank, uncomprehending stare."

Many growers I have spoken to consider that the growers Dom's size have the toughest job. They don't have the full-time crews and equipment of the larger growers, and they are too big to do all the work themselves. They have just enough acreage to require hiring out some work on a seasonal basis, leasing equipment, and purchasing sand in the winter. Dom does have his own sand, but not at this bog. He relies on his brother, a full-time high school history teacher, to help him throughout the summer and on weekends at other times of the year. In addition, he hires two part-time employees, plus seasonal workers during harvest and when needed for other jobs.

The Fernandes are one of the early Cape Verdean families who worked on the cranberry bogs when they first came to this country. The islands of Cape Verde supplied much of the early labor on bogs in the East. Most workers came from the agricultural island of Fogo after it had experienced several years of famine and drought. Boats, such as the schooner *Ernestina*, brought fresh immigrants from the Portuguese islands to New Bedford, where the cranberry grower Abel D. Makepeace was often waiting at the dock. After sizing them up, Makepeace would sign up the best labor prospects "right off the boat" and have them delivered to one of his bog shanties. There they found a communal pump and a basin beside each shack for the family to use for washing and cooking. Immigrant families often came as a unit. The mother in the family was instructed to prepare breakfast for her family, then pack a lunch for each family member. A truck would pick the family up around ten or eleven in the morning, after the dew was off the vines, and bring them back to the shanty around four or four thirty in the afternoon.

In Massachusetts child-labor laws prohibited hiring children under the age of fourteen, but many pickers were eight or younger, and the laws were largely unenforced. A 1911 photograph in the Spinner Collection at the Falmouth Historical Society shows Manual Rose, age nine, and Lena Rose, age seven, picking cranberries on the Swift family bogs alongside their mother and a crew of other workers. Margaret Fernandes, who worked on the bogs as a child, remembers setting out vines while on her knees with snow on the ground. Children who worked on the bogs often returned to school two months after the start of the school year and, having missed that much, had a hard time catching up.

In New Jersey, where there were virtually no child-labor laws until 1920, picking was done by hand without the use of the wooden hand scoops found on Massachusetts bogs, thus requiring more labor. It is estimated that as many as 1,200 children worked on the bogs from sunrise until sundown seven days a week. Despite efforts to improve working conditions for women and children, it took the Great Depression, when men sought any work they could get, to make bog work for Italian, Finnish, and Cape Verdean children obsolete.

Manual labor on cranberry bogs can be seen in several different ways, depending on the role and disposition of the viewer. According to author Joseph J. White, writing in 1912,

> The picking season is a pleasant one, for several reasons, to both picker and proprietor. The weather is proverbially fine in that most delightful of all months, October, when women and children turn out in great numbers to join the "cranberry picking" frolic, with well-filled dinner baskets and happy countenances.

The late Silvino "Ed" Fernandes (a man known for his sunny outlook) worked the cranberry bogs as a child with his family. As a twelve-year-old, he was paid twenty-five cents per hour. "We did everything by hand." That included hauling sand to build the bogs, scooping, and carrying boxes to load the harvested berries. He laughed as he admitted, "Do you want to know the worst of it? It was fun, like a game we were playing." Marilyn Halter, a historian on immigration studies, agrees that there were unexpected benefits to children. "They could be outdoors with their families, not separated from their mothers. And the racism found in textile mill work was not found on the bogs." The workday ended at dusk. Then out came the fiddles, guitars, and banjos, calling both children and parents to sing and dance in the fresh air. Years later Mary Andrade would reminisce about her days spent working on the bogs as a child: "They were happy memories in a way. Everybody told stories. It was a manner of socializing with our peers. We had a lot of fun on the bogs—as long as we kept on picking."

Some of the more industrious Cape Verdean families, thanks to the hard work of each family member, were able to save enough to buy their own cranberry bogs. Dom's father's family bought and built bogs in Duxbury, Massachusetts, and his mother's family bought and built bogs in Carver, including the Fresh Meadow bog, where we are today.

"I grew up in the cranberry business," Dom tells me. "I'm actually third generation, but if you had asked me thirty years ago if I wanted to grow cranberries, I would have answered, "Absolutely not." Dom's reentry into the business echoes that of Andrew Rinta and others. Four years after he graduated from college, his father became ill. At that time, his father was farming about ten acres of cranberry bogs by himself. Dom was about to switch jobs. He left his job, returned to help his family sort out what to do with the property, and enrolled in a graduate program where he earned an MBA. The year was 1983, 1984. Ocean Spray was paying fifty to sixty dollars per barrel for cranberries, more than what it pays for cranberries today, at a time when it was worth more. Dom's uncle, who owned adjacent bogs, wanted to sell. Dom put together a business plan, acquired his uncle's bogs, and started to work full-time growing cranberries. "And here I am, thirty years later," he says with a laugh.

From the mid-1980s to the late 1990s, cranberries had a good run from a profit point of view. "I could do as well raising cranberries in 1984 as I could as an entry-level MBA." Before that period cranberries provided a good quality of life for the growers, but it was a hard way to make a living.

The year 1999 was a watershed for growers hit with overproduction and not enough markets to absorb the excess fruit. Fifteen years earlier Dom and other growers had begun to remove some of the lower-producing indigenous vines, Early Blacks and Howes, and to replant the bogs with higher-yield Ben Lears and Stevens. "It was becoming more apparent in the industry," Dom says, "that you either grow high-yield varieties or you niche." One way to niche in the cranberry industry is to grow organically. These particular three acres had the potential to become an organic bog because they have their own source of water.

"The first year, it was an experiment." He chuckles. "And six years later, it's still an experiment."

I mention the many insects I see flying around and inquire how he protects his plants from them without using pesticides.

"Late water," he replies.

Growing organic cranberries while keeping the crop from being destroyed by pests is difficult to do, especially in New England, where a host of insects arrive at about the same time as the plant flowers. There is only one imperative cultural control, and that's "late water." The term refers to the process of reflooding a bog in late May and June, the period when insects arrive. With six inches of water covering the plants, bugs and other pests can't eat the plant or the berry. Fungus, if left untreated, will rot the berries, but the late water thins out the plants so that they don't form a canopy to create a moldy environment for fungus growth.

The downside to the late-water technique is that it lowers the yield. Growers who late water their nonorganic bogs often do it only every three years to give the yields a chance to improve. "How far it knocks down the yields, I don't know," Dom says, "but I'm going to find out because this is the fourth year I've done it."

Late water doesn't control weeds, "and that's the pest I'm struggling with because a lot of weeds can't be pulled out by hand once they've established roots," Dom explains. He is trying some organic vinegar on the weeds and careful burning of the seed, but the weed issue is ultimately going to determine whether or not this bog can remain organic. On the West Coast, a debilitating weed known as the lotus vine has destroyed whole cranberry farms.

It is much easier to grow organic fruit in Canada and, to some extent, in Washington, Oregon, and Wisconsin because of the climate differences. The warmer New England and New Jersey winters allow many insects to spend the winter in the bark of native trees. When the bugs emerge for the spring, they want to eat, and cranberries make a tasty meal. In Carver, Middleborough, Wareham, and Lakeville, Massachusetts, many of the growers share water aquifers. If more growers were utilizing the late-water technique, it would begin to put too much pressure on the water supply. "So," Dom predicts, "I don't see a huge future for organic fresh fruit from this region." That said, he points out the advantage of cooking with locally produced fresh fruit as opposed to fruit trucked from Canada or Wisconsin, where the berries are wet harvested, then dried, but "don't maintain the same keeping quality."

Dom is presently trying to work out the best balance between wet picking and dry picking the Fresh Meadow bog. He estimates a 30 percent loss of yield over several years of dry picking due to the damage the picking machines and pickers do to the bog during harvest. Last year, he only dry picked, sold the berries as fresh fruit, and barely broke even when he factored in the stress to the bog. This year he plans to dry pick enough to sell as organic fresh fruit at a makeshift stand beside the bog, at farmers' markets, and to several wholesale outlets. He hopes to sell at the stand only on weekends.

By the last day of September Dom has been able to dry pick for only two days. Rainy days have been predicted for the next weekend, and the question of whether to continue dry picking is still on the table. At the twenty-two acres where he grows conventional, non-organic berries and wet picks them, he and his crew have been working twelve-hour days, and the bogs have been yielding somewhere between 150 to 200 barrels per acre at a value of around forty-five dollars per barrel. The Fresh Meadow bog is expected to produce only around 70 barrels per acre, for a value of seventy dollars a barrel, offset by the stress to the bog and the stress to Dom in having to find markets for his organic fruit. Unless something changes, production of local organic fresh fruit in Massachusetts is like watching a swan start to lift off the water and wondering if it is worth the effort.

I ask Dom if he is concerned about large-scale, publicly held cranberry producers such as John Hancock, companies that may be operating with a different focus, that is, short-term profits to stockholders, as opposed to a long-term approach with a view to future generations. He points out that Ocean Spray, where he is a member, has been buying up and holding large tracts of land in Canada, possibly as a future hedge against the effects of climate change. "And if you are in the co-op, you kind of have to trust the decision making that is going on at the board level. We elect the board members, and the hope is that decisions that are made are going to be good for the majority of growers." He sees Ocean Spray as a safe home. When things are bad, he knows he has a market for his berries. For the past four years Ocean Spray has also been paying considerably more than has been paid to the independent growers. For Dom and others, that's another good reason to be part of the cooperative.

Farmers, ranchers, and fishermen have launched a food revolution in the United States. Within the past twenty years, they have used natural soils and waters, pastures and grains to reach incomparable flavors in food . . . What Americans have done is harvest excellence.

SUSTAINABLE

Oregon's south coast is home to some of the last and healthiest wild salmon south of Alaska. They swim and spawn alongside more than two hundred cranberry farms carved out of what was some of Oregon's least productive agricultural land. Previously, efforts to farm cranberries next to the river and stream habitats for trout and other spawning fish would have been loudly condemned by the environmental community. Today, as we learn more about the properties of bogs to sequester CO_2 from the atmosphere, mutual coexistence is encouraged.

Since 1874 the McKenzie family has been raising sheep, cattle, and salmon on close to two thousand acres atop the westernmost point of land on the Oregon coast. "Every place I go on the farm, there's a memory of my father, . . . my grandfather," Scott McKenzie says with fondness. For the past forty years, the family has also grown cranberries. Cranberries are in the family's veins. Scott's wife, Carol Bates McKenzie, is the granddaughter of the man considered the pioneer of cranberry growth in the Coos Bay and Bandon area. During the Depression Ray Bates and his brother planted two acres of cranberry beds by hand. One acre was planted with Stankavitch, a hybrid named after the two Wisconsin brothers who created it by crossing a wild Oregon berry with an eastern berry. The other used mowings of the McFarlin variety, the berry Charles McFarlin brought to the West Coast from his Carver, Massachusetts, bogs in 1885. McFarlins are still grown on two cranberry farms I visited in Coos County. The Stankavitch vines appear to have been replaced with higher-yielding Stevens.

Scott and Carol devote eighty acres to growing cranberries. They presently harvest forty-five acres of mostly Stevens along with a high-producing variety called Yellow River, genetically grown on bogs in the Yellow River area of Wisconsin. The McKenzie cranberry vines grow in six to eight inches of sand layered over clay to prevent water from leaching out when the bogs are flooded. By keeping the planting medium wet and cultivating a plant that takes in CO_2, Scott claims that the absorption of CO_2 is similar to that of eastern peat bogs where carbon is sequestered because organic carbon doesn't oxidize easily. As the plant grows, it absorbs carbon dioxide from the air and the carbon stays in the peat.

For the past three years, Scott and Carol have been growing cranberries sustainably. The United Nations defines "sustainability" as "meet[ing] the needs of the present without compromising the ability of future generations to meet their own needs." Scott proudly describes Clearwater Cranberries, a joint venture between the McKenzie Seaview Farm and Randy and Gretchen Farr's Elk River Farms, as a collaboration whose mission is to market a cranberry grown sustainably, with a commitment to protecting the environment and promoting local farm-grown produce. It is a tall order. Clearwater's owners substitute plant oils for toxic pesticides, they practice late flooding and sanding, they regularly monitor bog health on the premise that a healthy vine can best ward off insects and other pests, they fertilize with compost tea and fish oils, and they recycle water to prevent it from running back into the rivers and streams of the local watershed.

As a result of their environmentally sensitive cultivation practices,

Clearwater Cranberries has earned Food Alliance certification, been cited as Watershed Friendly Stewards by the Southwest Oregon Resource Conservation and Development Council, and is certified as "Salmon Safe" by Oregon's Salmon-Safe, a nonprofit working to "keep the urban and agricultural watersheds clean enough for native salmon to spawn and thrive." The Farrs and McKenzies hope their environmental efforts will help create a brand name sought after by buyers.

Below the cliff where Scott McKenzie's and his mother's homes sit, waves roll off the Pacific to crash on the beach. Cougar, bobcat, elk, and deer roam the property. A stream runs through and empties into the Pacific. The Clearwater cranberry farm of Scott's partners sits on one of the finest salmon streams in the state. Together, the two families have relied on the land to provide for them and their families. "Now it's time for us to take care of the land," Scott says. "We're trying to sell cranberries as close to where we're producing as possible, to use as few outside chemicals as possible while still maintaining our yields, and to protect the quality of the soil and water as much as possible." He points out that Clearwater Farms's berries are considered sustainable, not certified organic.

Scott and Carol considered a different form of cranberry growing in 2007 when they first learned that outsiders were trying to buy local land as a destination for golf courses. At that time the McKenzies saw the price of farmland escalating. In response they joined neighbors to find a means where local farmers could make a good enough living so they wouldn't be forced to sell off their land. Both the U.S. Fish and Wildlife Foundation and the Kellogg Foundation offered grants for the purpose of keeping farmers on their land. Thanks to the efforts of the McKenzies, Farrs, and others, cranberries were the first product the foundation decided to fund on Oregon's south coast.

A West Coast grower's greatest monetary cost is for irrigation. On the East Coast, rainfall has traditionally averaged fifty inches per year, evenly scattered throughout the twelve months. Rainfall on the West Coast traditionally averages one hundred inches in the winter, with four to six months without rain in the summer, hence the fragile nature of the drainage basin during the summer months. The McKenzies and Farrs had established a reputation for working to protect the local Oregon watershed, making them the likely choice for funding.

Prior to receiving the grant, Scott had been selling his berries on his own as quick-frozen berries and as concentrate that he sold to stores who wanted to use their own store labels. The grant allowed the two families to hire a marketing firm to study whether grocery stores, bakeries, food services, restaurants, and distributors would pay more for regional cranberries that were sustainably grown by small independent farmers.

Most of the respondents to the marketing study indicated that they would like to purchase sweetened dried cranberries that were sustainably grown. On the basis of the study, the Farrs and McKenzies convinced the Scenic Fruit company, the firm that quick-froze Scott's cranberries, to fire up a drying system the company had "on the drawing board." After running a few experimental batches, Scenic Fruit determined that, for them, drying cranberries wasn't economically feasible. Scott turned to an Oregon processing plant that had been previously certified by the Food Alliance to process cherries sustainably. The result: in addition to its frozen sliced berries and frozen whole berries, Clearwater Farms has just sold its first sustainably grown dried cranberries. Best of all, they're preserving the land and habitat for their family and the creatures sharing it with them. Hopefully, their efforts can enable the salmon and steelhead to continue to spawn in the Elk River and its tributaries.

Nowhere on the American continent except in the choicest irrigated districts, has wild and apparently worthless land been taken from the state of nature and developed and made so valuable.

ALMOST CONVENTIONAL

In Warrens, Wisconsin (population 373), it's possible to choose from five different flavors of cranberry ice cream, plus a cranberry sherbet. There's straight cranberry, cranberry truffle, cranberry cheesecake, chocolate cranberry caramel swirl, and—for the true chocolate and cranberry lover—rockin' choco berry. It's all available at the Wisconsin Cranberry Discovery Center, a museum with its own ice cream parlor housed in a former cranberry warehouse. I chose the "rockin' choco" and was rewarded with a rich cranberry sauce swirled into chocolate ice cream and dotted with chocolate-covered dried cranberries. I want to go back for more.

Down the road from the Discovery Center, Nodji Van Wychen, a member of the center's advisory board, grows cranberries on 110 acres of wetlands next to her family home. One-third of the acreage is devoted to packaged fresh fruit sold under their Wetherby label and their Crane Trail label; the other two-thirds is sold for juice. They design and make most of their own equipment, and family members do most of the work.

As elsewhere, Wisconsin cranberry history begins with the natives. The Potawatomi, the Menominee, the Chippewa, and the Winnebago gathered the low swamp berries they called *atoca*, then traded them with French Canadian fur traders for beads, knives, whiskey, and flour. Until 1850 the tribes picked the wild berries for free. After that, treaties with the U.S. government, including the so-called swampland grant, transferred the cranberry swamps to white settlers who began to buy up land and harvest the wild berries.

The confiscation of the cranberry swampland conveniently dovetailed with the expansion of the railway west beyond Chicago. Easterners headed west to make their fortunes, and many took note of the get-rich-quick potential of the tart, little berry they had previously connected with Cape Cod. Edward Sacket, one of the out-of-state cranberry pioneers, left Sackets Harbor, New York, to buy a cranberry swamp in Aurora, Wisconsin. He lost his first crop to flooding followed by drought. Undaunted, he cleared the land, dug ditches, and built dams before planting his next crop of cranberries. The result: in 1865 he sold a thousand barrels of cranberries at fifteen dollars a barrel. Wisconsin cranberry cultivation was off and running. Chicago investors saw a commodity that could provide short-term profits for a relatively small investment, and they began to put together joint ventures. The Wetherby marsh was one of them.

Nodji's grandfather, Henry Kissinger, originally worked as a foreman for the Wetherby Cranberry Company, which was named for one of its original 1903 Chicago investors. In lean years Wetherby paid its foreman in stock instead of cash. When Nodji's grandfather retired, he held forty shares of stock out of a total of five hundred, and he passed them along to his daughter, Nodji's mother. By 1972 Nodji's mother and father, both teachers, had saved enough from their salaries to buy the remaining shares, thus making the Kissinger–Van Wychen family sole owners of the Wetherby marsh. Recently, Nodji and her husband, Jim, converted sixty acres of uplands for their son and son-in-law. They installed PVC piping and tiles, similar to those installed by Keith Mann in Massachusetts. They upgraded their pump house. And they automated their packing and sorting operations.

For over 150 years cranberry sorting has been based on the premise that a ripe cranberry will bounce. The first man credited with this discovery is said to have been one of the early New Jersey growers, John "Peg Leg" Webb, who was left with one leg after a tree fell on the other. John stored his berries on the second floor of his barn and one day he is reported to have fallen down the stairs while carrying a barrel of cranberries. Ever observant, he noticed that the ripe berries bounced and the bad ones either stayed on the steps or rolled to the side. Since that time cranberry separators have been variations on those stairs, consisting of wooden slats or "bounce boards" that allow the good berries to bounce into one bin and the soft or damaged berries to fall into another. It is probably no coincidence that cranberry separators were first produced in New Jersey, John's home state. Twelve years later, in 1882, the Hayden separator was designed and built in Carver, Massachusetts. The Bailey separator wasn't patented until forty-one years later, and, though no longer made, is still used at Keith Mann's organic screening operation in Buzzards Bay, Massachusetts.

The U.S. government recently decided that wood and food do not mix. The result: good management practices are phasing out separators with wooden bounce boards. For growers of fresh, whole cranberries in Wisconsin, that posed a challenge. Nodji and her family decided to automate. After an expenditure of half a million dollars, they now have the capability to sort, process, and package ten thousand pounds of cranberries per hour, five times more than before automation.

"Before our new equipment began operating, I was so proud of being able to use the sorting equipment my grandfather used in 1920," Nodji says, "and it was very efficient." But, bounce machines are not perfect and growers have traditionally relied on experienced hands and eyes to spot the unacceptable berry that successfully makes it to a sorting table. In a town the size of Warrens, many of those people have been lost to attrition, and young people aren't interested in acquiring the training to replace them. Two women are all that remain. They have been hand sorting cranberries at the Wetherby marsh for close to fifty years and are still performing the final hand sorting of the berries that make it past the new system's lasers, cameras, scanners, and other computer-driven equipment.

Less than 5 percent of the cranberries grown in the United States is harvested and packed as fresh berries. The remaining 95 percent is sold either for juice or sweetened dried berries for breakfast cereals, snacks, or trail mix. During the harvest season, anyone wishing to purchase fresh Wisconsin berries can stop at the edge of the road, walk to the Wetherby warehouse, and buy cranberries, often from Nodji or another Van Wychen family member. A home cook can buy a one-pound bag with the Wetherby logo or a twelve-ounce Crane Trail bag artfully printed with recipes and a drawing of a sandhill crane. Or the consumer can purchase a five-pound or thirty-pound bulk box to freeze berries to use throughout the year. The family's fresh berries are also available at the Dane County market, a farmers' market about a hundred miles south in Madison. Strict regulations stipulate that a farmer must own more than 50 percent of her operation to sell at the market. Nodji is the market's only cranberry vendor. And she likes it that way.

"We're not your typical grower," Nodji explains, "because we have so much of our harvest in fresh fruit, and we work with the public." She adds that most growers prefer to align themselves with a large cooperative such as Ocean Spray and "once the berries leave their marsh, the growers turn the responsibility over to someone else."

Typical growers also don't produce a line of cranberry wines. In 1999, before sweetened dried berries were readily available and before the foreign markets had been developed, cranberry production exceeded the demand. Many growers cite that year as a tipping point for commercial cranberries. Growers were forced to sell their berries for less than it cost to grow them, and the Cranberry Marketing Committee demanded that its member-growers dump berries instead of selling them. Jim Van Wychen didn't want to throw away good cranberries. In a variation of lemons into lemonade, Nodji and Jim, instead of destroying their crop, turned it into wine. The result was the birth of the Van Wychen wine company.

Recently passed Wisconsin legislation stipulates that all wine producers under a certain size, including Wetherby, must sell through a distributor. Jim Van Wychen's wine production was so small that he wasn't able to interest a distributor in taking on the line. He and Nodji appealed to the legislators and finally received permission from the Wisconsin Department of Agriculture to form a cooperative distributorship. Only one other small wine maker exists in the state, but as of June 30, 2009, the two tiny vintners are allowed to distribute their wines through their own cooperative.

Unlike grape wines, cranberry wines don't benefit from being laid away in a cellar to age, and cranberry wine blends have an even shorter shelf life. After six months they go bad. Nodji can't stand having the Wetherby name associated with a product that isn't perfect, even though it may have been perfect when sold to the customer, so a cranberry-apple wine blend was discontinued. Now the family offers only the undiluted variety.

"We've been packing fresh fruit on this marsh since 1905, and we've always taken so much pride in producing a high-quality product," Nodji says. "The Wetherby name on our bags has come to be respected over the years, and that's because our family owns it and we see every bag that goes out of our facility." She adds that Byerlys, a "very upscale" grocery store throughout the Minneapolis–Saint Paul area, has sold only the Wetherby line of cranberries since 1936.

Listening to Nodji, I can't imagine her turning her back on an opportunity to educate the public, especially when the subject is something she is as passionate about as cranberries. She leads bus tours of the marsh during blossom time and harvest, she speaks to school groups, and each year on the first Saturday in October she leads a cranberry harvest day, an educational day for the public to learn about cranberry production.

Wetherby and a few other Wisconsin marshes employ a unique variation on the harvest process. The marsh is flooded. A mechanical raking machine at the edge of the marsh scoops up the berries and feeds them onto a conveyor belt, where they are gently deposited into a series of floating boats. When full, the boats are moved to the end of the bed. A specially designed clamp on a hydraulic lift picks up the boats and moves them to a waiting truck heading to the Wetherby warehouse. There, the berries are dried and stored until sorted and packaged. Unlike eastern harvests, Wisconsin does no dry harvesting—even for whole, fresh berries.

To be available to the customer throughout the year, fresh cranberries require cold storage. Most markets don't want to devote space and money to store cranberries, processors don't want to purchase or rent storage for cranberries, and growers can't justify the expense of storing fresh berries. The result is that the home cook who wants to cook with cranberries throughout the year has to buy in bulk during harvest, then free up enough home freezer space to preserve the fruit. Those who know this buy in bulk at the Dane County Fair or the Warren Cranberry Festival or from the Wetherby warehouse on the marsh. For them, a stock of cranberries that will last from harvest to harvest is well worth the drive from Illinois, from Ohio, from Minnesota, from Missouri. "Harvest is a pretty time of the year for a drive, and," Nodji admits, "it helps reduce the number of cranberries I need to store." She and her family maintain a ten-thousand-cubic-foot freezer for storing fresh berries, but that is not enough to handle the harvest. To help ease the pressure, the Van Wychens offer to ship berries directly to the consumer. They're one of the few growers who do.

One cranberry variety grown on the Wetherby marsh is the same variety originally cultivated by Eli Howes on Cape Cod. Few other Wisconsin growers grow this berry because, as Dom Fernandes noted, the yield isn't as high as some of the hybrids. "But," Nodji points out, "the size is uniform. You don't tend to get little berries and bigger berries in the same bag. And they have a thick, waxy but shiny skin." The skin texture makes them hardy and allows the berry to hold up well in a grocery store, where the produce manager may not handle them with the tender loving care many more fragile berries require.

We are seated in Nodji's family home, around the corner from her mother's house. She shares a driveway with the family of one of her children, and grandchildren's toys are scattered at the edge of the drive. Inside, on either side of a local pudding-stone fireplace, long windows enable us to look out over fields of cranberry marshes, fringed with stands of pine and bisected with thin strips of water. Bees pollinated the fruit over a month ago, and now it is up to the sun to ripen the buds and the rain to keep the vines moist.

One can no more approach people without love than one can approach bees without care. Such is the quality of bees ...

—Leo Tolstoy, beekeeper

7

POLLINATION

They sailed to the Western Sea, they did,
To a land all covered with trees,
And they bought an Owl, and a useful Cart,
And a pound of rice, and a Cranberry Tart,
And a hive of silvery bees,

—Edward Lear, *Jumblies*, 1968

Two-thirds of the three million bee colonies in the United States are transported to pollinate the crops we depend on for food. When David Mendes and his Headwaters Farm flatbed trucks roll up and down the highways, they are transporting close to four million honeybees stacked in wooden hives protected by a thin mesh screen. Honeybees don't like to be moved and will tolerate it only if lulled by the motion of the moving truck. If the truck slows down, the bees get angry. And no one wants to deal with four million angry bees, especially since 2006, when countless bees began to die and the evocative term, "colony collapse disorder" (CCD) was subsequently coined.

Plausible causes for the collapse range from malnutrition to debilitating mites and just about everything in between. To date, no one involved knows the full cause, only pieces of the puzzle. A majority of apiarists and scientists appear to agree that bee deaths are probably caused by a combination of factors falling into three general categories: pesticides, pathogens, and poor nutrition due to habitat loss. Scientists focus on pesticides as the primary cause, chemical companies focus on pathogens, and everyone agrees that the bees are malnourished, either from habitat loss or an inability to secure a balanced diet. The latter is due in large part to the conversion of farmlands from a more diverse mix of crops to corn and soybeans—especially since the development of a larger market for ethanol.

The pesticides that have received the most publicity with regards to bee deaths are imidacloprid and its successor clothiniadin, both part of a group of insecticides known as neonicotinoids. Imidacloprid has been approved in the United States since the 1990s, partly because it was believed to be nontoxic to mammals. Its use as a seed treatment was banned in France in 1999 and in Germany and Italy in 2008. In Italy the 2009 test crops that remained neonicotinoid-free showed no massive bee mortalities, a phenomenon not seen since 1999.

On April 29, 2013, the large numbers of deaths worldwide prompted the European Commission to institute a two-year temporary ban on the use of three neonicotinoids to ascertain if the mortality rates for bees would decline within the test period. As of December 2013, despite the death in June of twenty-five thousand bees in Oregon and despite at least one pending lawsuit from beekeepers and environmental groups for failure to protect the bees, the U.S. Environmental Agency has taken no action to ban neonicotinoids.

"They keep telling us that bee scientists have to prove to the EPA that there's a problem," comments David Hackenberg, the beekeeper credited with first noting Colony Collapse Disorder. "The problem is that the EPA is supposed to protect the environment. It's their responsibility to make sure that the chemical companies are doing their job."

Susan Kegley is CEO and founder of the Pesticide Research Institute. When we met, she showed me the results of a survey of beekeepers her firm had conducted to try to better understand what crops might be harboring the greatest amount of pesticides responsible for causing the greatest number of bee fatalities. She discovered, among other factors, that government inspectors were not inspecting beekeepers and concluded that since their job usually consisted of inspecting cattle feedlots or poultry farms for substandard conditions, they just weren't geared to inspect hives of insects—especially migrant insects.

To my dismay, I discovered that the Bayer AG product containing imidacloprid is, in 2014, readily available at most gardening supply stores and that it is being bought and used by home gardeners despite the warning label saying that it kills bees. Imidacloprid is not generally used on cranberry bogs, but pollen sampling is indicating that chemicals targeted for certain crops and chemicals used on lawns have blown onto the weeds and wildflowers preferred by foraging bees. As early as 2007 David began to question whether or not fungicides could be harmful to his bees. Fungicides used on cranberry bogs work by coating the flower so that fungus can't permeate the barrier. Unlike pesticides, fungicides have not been designed to harm insects, and it has been generally accepted that fungicides were not harmful to bees and that honeybees could safely be working on the bogs in the window of time between the application of fungicides and the application of pesticides.

In July 2013 a study was released describing the findings of a group of researchers who collected and analyzed pollen samples from honeybee hives that had been delivered to pollinate crops, including cranberries. Samples showed that the pollen collected contained both pesticides and fungicides. This opened up the possibility that if the bees were somehow bringing fungicides into the hives, the possibility of damage to the bees from fungicides at least deserved more attention. The researchers also determined that not all the pollen gathered came from the targeted crops, but that the bees had also gathered pesticides and fungicides while foraging on the weeds and wildflowers at the edges of the fields and bogs. In the hives near cranberry bogs, more fungicides and pesticides seemed to come from the plants surrounding the bog, than from the cranberry plants, possibly due to the use of integrated pest management versus general spraying.

Without bee pollination U.S. fruit crops would decrease by approximately $15 billion. Without bees New England's cranberry, blueberry, and apple markets, estimated at $121 million per year, would cease to exist. Timing is essential for successful propagation. Pollination has to be synchronized with plant growth or the plants won't produce berries. Right now, Dom Fernandes's bogs have a light, velvety texture, a result of the many blossoms waiting to be pollinated. "These hives were delivered here two weeks ago," he tells me, "and the bees have been on the bog for no more than a day and a half."

In a normal year in late June, honeybees would be pollinating the flowers and the fruit would have begun to set. Fortunately, Dom's Fresh Meadow bog has enough surrounding vegetation to attract and hold a large population of native bumblebees. In 2011 the *Boston Globe* reported that the last time the month of June saw fewer hours of sun was in 1903. Given a spring and early summer that have seen day after day of driving rain, even the bumblebees aren't working. And Dom is concerned. "I'm very nervous this year because it has just been such an abnormal year. We're all hoping this weather pattern breaks. And breaks fast," he stresses.

Dom's organic Fresh Meadow bog may have an advantage over conventional bogs because its plants are still in the "hook," or preflowering stage, where the bud has started to resemble a crane's head, but the flowers have yet to blossom. Late water, the process employed on most organic bogs, slows down flowering. Across the bog two beehives sit in the wet grass. No bees are visible on the bog. By the time Dom's organic plants flower, the rain hopefully will have stopped, the plants will be ready for pollination, and the bees will be at work.

Honeybees are not native to North America. Honey was the sweetener most commonly used by the English common folk in the seventeenth century. Only the aristocracy could afford sugar. Even after the sugarcane plantations were established in Barbados in the seventeenth century, English law decreed that raw sugar had to be shipped to England for processing. To ensure a supply of honey in the colonies, early settlers imported German Blacks (*Apis mellifera mellifera*), a bee known for its hardiness and longer lifespan and for not swarming as much as other strains. The first imports are believed to have been brought to Virginia in 1622. Over time, many of the imported bees chose to settle in the wild. There are approximately 17,000 known species of bees throughout the world, 4,000 in North America. Some are smaller than a mosquito. Just about all of them are more efficient pollinators than the honeybee. It takes approximately 100,000 honeybees to pollinate the same amount of flowers as it takes 500 mason bees (genus *Osmia*). It takes 3,000 honeybees to do the work of only 150 leaf cutter bees (genus *Megachile*), another native pollinator. Unlike honeybees, native bumblebees (genus *Bombus*), mason bees, and leaf cutter bees work in high winds. They normally work in rain. They work in colder weather. They're harder workers. So why rely on honeybees? Across town, the Rinta family are seriously addressing that question.

A thrush warbles somewhere in the field as I walk toward a blueberry patch adjacent to Andrew Rinta's family's bogs. The blueberry bushes are in flower, drooping creamy bells presaging future berries. For over thirty years, the Rinta family have cultivated these blueberries. When the blueberries are ripe, the family opens a pay-as-you-pick stand between the blueberry bushes and their road. "My friends and I all did 'patch duty' here when we were kids. We've all worked in the stand making change," Andrew reminisces.

This year the blueberries are part of a bee experiment. Up to now, the Rintas have relied on David Mendes's honeybees to pollinate their blueberry bushes and then the cranberry vines. This has worked well because the blueberries are in flower about a week or two before the "crane head" of the cranberry bends, indicating that the plant is ripe for pollination.

Four cardboard, biodegradable bumblebee colonies have just been delivered to the Rinta bogs, where they sit at the edge of the blueberry patch. Each of David Mendes's honeybee hives holds a colony of forty thousand to sixty thousand bees. Each box of bumblebees contains fewer bees, but the bumblebees are expected to be more efficient than the honeybees.

For the grower, bumblebees have another advantage. They can be moved and will return to the hive even if it isn't in the same spot as when they left. Unlike honeybees, they can be directed toward the plants a grower would like to see pollinated. There is just one problem: "There isn't enough native food to keep the bumblebees around once they've worked the cranberry bogs," Linda Rinta, Andrew's mother, explains. She has a possible solution. Linda would like to see beds of red clover (*Trifolium pratense*), and other forage plants, planted under power lines and other utility corridors. The pollinator gardens, with their ability to offer the bees pollen and nectar for a longer period of time, would presumably protect both the flowers and the bees from residential and commercial development while providing a well-balanced diet for the bees. Essentially, she would create bee parks in a nonutilized area.

More and more farmers are recognizing that maintaining the health of the bee populations is critical to maintaining the health of the farm. And the focus on bees is not coming too soon. The decline in honeybee populations has been well publicized. Less well known is that some bumblebee populations have declined as much as 96 percent in the past twenty years, while their ranges have diminished by 87 percent. The population for certain types of bees has sunk to a level that qualifies them for the U.S. Endangered Species List. Unquestionably, pesticides have contributed to the decline, but bees need to eat, and if the nectars they feed on no longer exist because of environmental degradation, drought, or other results of a warming planet, they starve. In addition, as trees and other habitats that afforded wild bees shelter are being destroyed, the bees are left out in the cold to freeze. Imagine leaving our ability to feed ourselves and others in the hands of someone who is homeless, with no protection and no assurance of a meal.

Until recently, bumblebees were confined to greenhouses. Due to the cost, a grower couldn't afford to purchase a bumblebee hive and just put it out for the bumblebees to vacate after pollinating one crop of cranberries. Under Linda's plan, a habitat would be created to maintain a sustainable population of bumblebees. Once they work the cranberry bogs, the bumblebees would move on to pollinate the clover and various late-blooming flowers under the high wires. When the time came for the queens to hibernate, they could seek out one of the many hollow plant stems or crawl under the bark of nearby trees until needed to establish next year's colony in the wild and pollinate another year's crop of cranberries and blueberries. Of course, this plan is predicated on the assumption that the balance of nature will not change, and few climate scientists see that as a strong possibility.

Bumblebees on the Rinta property are now busily buzzing around the blueberries. Cranberries flower just after blueberries. When the cranberries are in flower, Andrew will move the hives closer to the cranberries and, as he says, "point the way to the bogs." Just in case he needs a backup or reinforcement, he has placed an order for honeybees. At this point it's an experiment.

The day after my visit with Dom Fernandes, the sun began to burn through the clouds. At Keith Mann's bogs, bees were plentiful. "They are really working," Keith says. He also expresses concern that unless the sunny days continue, the berries will suffer from fruit rot and other insect-related problems. Pollination must take place while the blossoms are ripe, but that is also the time when the fruitworm, commonly known as "spag," arrives and fungicides need to be applied on the bogs that weren't late watered. It is also just before the time that various other cranberry pests attack and conventional growers apply pesticides to ward off the insects. Keith Mann tried some bumblebees two seasons ago. "They cost a fortune," he comments, "and you have to order them long in advance because they [bumblebee keepers] grow the bees for each order."

The busy-bee image is better suited to bumblebees than to the more persnickety honeybee. Bumblebees are most active in the morning and early evening; honeybees like to sleep late. Ideally, a grower would have some of each type of bee on the bogs. "But," Keith says, "I had my scouts out there looking for insects, and they said they never saw a bumblebee on the bog." I offer the possibility that maybe the bumblebees work at night, which Keith finds amusing. Despite the lack of bumblebee sightings, it was still the best crop ever. Keith attributes the success of the pollination to David Mendes's honeybees trucked from Florida.

Transporting bees is not a modern-day phenomenon. In 256 BCE, during the Ptolemaic period, a beekeeper named Senchon petitioned to have her donkey returned to her so that she could move her hives from the fields to higher ground before the Nile flooded and destroyed them. Today David Mendes and the 1,300 other migratory U.S. beekeepers arrive and depart from orchards, fields, and bogs within that short window of time from when the plants are in bloom to the time when pesticides are applied. Carolyn DeMoranville at the Experiment Station sees it as a "juggling act," and one of the reasons the fit with migratory beekeepers like David has been beneficial.

In a sense, David Mendes is the equivalent of a Western cowboy, rounding up and transporting his livestock to New Jersey, Maine, and Massachusetts from Fort Myers, Florida, via the highways. When he arrives each year, he parks his trucks in the field he purchased on the Wareham-Carver line to give him a location close to the cranberry bogs for the short season between flowering and the arrival of less beneficial insects. His job has become far more stressful than it used to be. Each year he and the other migrant beekeepers have had to rebuild a large number of their colonies, because so many bees began mysteriously to die. They have been able to split the remaining healthy hives, but they have had to purchase additional queens to populate them, adding significantly to their overall costs, and they can pass only so much of that expense along to frugal growers, who are also struggling.

Just before he left to head south, I watched him smoking his bees to get them to return to the hives for transport. It was an eerie ballet, human figures clothed in protective gear swirling around swarms of bees—all enveloped in smoke. It often wasn't clear who was in control, the bees or the beekeepers. The next day, they were all gone—up in smoke until next year.

Over the winter, growers will be rehashing whether to order honeybees or bumblebees and, if so, how many of each. For the Rinta family, their experiment with commercial bumblebees convinced Linda that "the best pollinator is the native species of bumblebee, and our best bet is to enhance our environment to protect and encourage their healthy populations"—not a small task given our fragmented landscape, pesticide use, and the many variables apt to result from a warming climate.

8

TRICKING THE VINES

Dead sand, water and air, are the elements upon which the cran-berry feeds the best, and attains its highest degree of perfection.

—Benjamin Eastwood, *Cranberry and Its Culture*, 1847

Vines begin putting out new runners in July at the same time the flowers bloom on the previous year's growth. Every two to five years, a grower spreads one-half to three-quarters of an inch of sand on the bogs during the winter months. If a covering of sand is not ap-plied, one-foot to six-foot-long runners grow parallel to the peaty soil. The result: instead of producing berries on upright shoots, plants put forth buds only at the end of each runner. A layer of sand causes the runners to send vertical shoots upright to capture the sun's warmth. These uprights form the terminal buds out of which the flowers and their berries emerge.

Farm manager Tom Larrabee on Nantucket opines that "nutri-ents in the sand awaken the berry from dormancy." When the plant has accumulated enough chill hours (1,700 to 2,000 hours between thirty-two and forty-five degrees Fahrenheit) to store energy and complete the cycle, it will break dormancy and begin a new growth spurt. In the absence of this dormant period, as the planet warms, the cranberry plant may not be able to produce spring buds, summer flowers, and fall berries. I visited the Oiva Hannula bogs in Carver, Massachusetts, on a Friday in February to watch nine men, five sanders, one screener, two dump trucks, and two front-end loaders place sand on the last nine bogs to be sanded this season. Scott Hannula is a fourth-generation cranberry grower, proud descendant of the men and women who immigrated to the bogs because it reminded them of their native Finland, and there was work to be had. The Finnish word for swamp is *suomi*, and many a Finnish immigrant used the word to describe his or her homeland, a large part of it swamps and lakes abutting wooded areas, similar to most areas where cranberries are grown in the United States.

Around the mid-1880s Abel D. Makepeace, a highly respected grower and one of the founders of the Cape Cod Cranberry Growers' Association, was constructing 160 acres of bogs in Wareham and needed help. He had to look no further than the Finnish employees in the Makepeace West Barnstable Brick Company. Makepeace built many more bogs in both Massachusetts and New Jersey, and the work he provided drew more Finns to the bogs. By 1900, when the Russians were drafting Finns for the army, many Finnish men looked at the alternative of working on the U.S. cranberry bogs and decided to emigrate.

Scott Hannula recalls regularly driving the bogs as a boy with Oiva, his Finnish grandfather, and never wanting to do anything else after that except grow cranberries. Today, he manages 250 acres of producing bogs together with the surrounding upland acreage. It's a vertical operation. Very little needs to be purchased from outside. The Hannulas build and repair many of their own machines, control their water supply, own their own land including their sand and gravel operation, and employ eleven men full-time. All full-time employees live within eight miles of the Oiva Hannula bogs. One man, Donald Hitchcock, has worked for the company for over thirty years, and part-time pickers are brought in to help at harvest.

Yesterday, the temperature registered zero degrees Fahrenheit, with a steady wind that averaged seventeen knots. The sanding operation was rescheduled for today because "I want to treat my men better than asking them to go out in open-cab sanders on a day like this," Scott Hannula explains with an embarrassed laugh. Today, the wind has died down. At seven thirty in the morning the temperature reads five degrees. By ten thirty it rises to nineteen. The men want to get started so they can finish the job before the sun sets and the temperature drops again. They have been sanding seven days a week since January 7, and Scott has promised them a week's vacation once they've finished these last nine bogs.

We drive the older of the dump trucks to an area between a cranberry bog and a cliff, supporting the magnificent hemlocks and tall-trunk pines found throughout Carver. Below the cliff a front-end loader is depositing freshly dug yellow sand into one end of an elongated green-and-yellow machine. Inside, the sand is tumbling through a revolving series of canisters that screen out the rocks and debris. The clean, screened sand is then fed to a conveyer belt that carries it up to the level of a dump truck, then spews it into the box of the truck. Any boulders separated out will be ground up or sold for making pudding-stone walls. Little is wasted. Pit sand, usually from sandstone, must be screened and washed before being placed on the bogs. Optimum sand size and cleanliness allows the correct amount of drainage to take place and eliminates weed buildup. As each load of sand is dumped into its box, the dump truck lurches. When the box is full, Scott pulls the cover over the sand and we drive off.

On the bog the morning snow has been cleared. Yellow sanding jitneys, about the size and shape of golf carts, flit over the frozen ice, painting it a sand color, quadrant by quadrant. The truck backs up to an area between two bogs, and Scott presses the lever to release the sand. Once the box is empty, he drives back for a fresh supply. A front-end loader immediately digs into the pile of screened sand and deposits a shovel full through the grated area at the back of the first sander lined up to receive it. The sander darts away. It zips down an incline onto the bog and reaches the line where the last sander left off. The driver, slathered in a face mask, tightly wound scarf, and heavy gloves, pulls a lever to release a steady stream of sand to the bog. Each sander makes one and a half to two loops on the bog before it runs out of sand, then goes back up the ramp for a refill.

I ask the driver of an older sander with SPLASH written on the side how he knows when he's put the correct amount on the bog. "You measure it," he replies simply. "You get out and measure it by hand." Depending on the height of the uprights, some bogs will require a half inch of sand; others, three-quarters." Before the bogs were flooded, and before the ice froze, the men had calculated the height of the uprights on each bog. Then they waited for the ice to freeze solid enough for the sanders to drive over it. In early spring, when the ice begins to thaw, the sand will slowly filter down to the runners to encourage next year's upright growth.

Scott has perfectly timed the sanding operation. Five days later the temperature is fifty-nine degrees at noon. Insects have emerged from their winter homes to swarm in the warm sun. When the sun goes down, they'll return to the woods to take up lodging between the layers of bark and trunks of the nearby trees until spring, when the cranberry's new flowers provide sustenance. For now, colder days are expected.

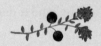

People don't realize how hard the farmer works. Too many think that all they need to do is go to the store and buy whatever they want. They think that food is permanently stored in cans in the cellar.

—Brud Phillips, cranberry grower, in
Robert Bartlett, *My Corner of New
England*, 1984

9

THE HARVEST, DRY OR WET?

The cranberry submits to cultivation, but it retains that savour of wildness that is its birthright. It cherishes that as a cat its spirit of independence. There are little things as well as great that cannot be tamed, cranberries and cats, as well as winds and waters.

—Cornelius Weygandt, *Down Jersey*, 1940

If sand provided enough nutrients in January . . . if frost was kept at bay in March . . . if late watering or chemicals licked the cranberry rot in May . . . if bees successfully pollinated the flowers in July, then ripe, red cranberries might be ready to harvest from September through December, depending on where they were grown. If . . .

Two weeks after the Piney Woods harvest, four men move behind picking machines on an organic bog in Wareham, Massachusetts. The man wearing a short-sleeved loose-fitting T-shirt is Keith Mann. All the berries grown on these bogs will be sold as fresh fruit, cleaned, sorted, bagged, labeled, and packed for shipment at a 1930s screen house on the Mann family's Buzzards Bay property. Little to nothing is farmed out.

Dry picking is far more labor-intensive and time consuming than flooding the bogs, floating the berries to the surface, and corralling them into a chute. It will take nine people three days to harvest these seven to nine acres of organically grown berries. It took part of a day and four to five men to wet harvest the Piney Woods bog, half the size of these bogs. So why do it?

For Keith Mann and his wife, Monika, it is a state of mind. They eat only organically grown cranberries and want to sell organic berries to the public. A line drawing of the two of them with their daughter is printed on the bags labeled with their ORCRANICS logo. On the back side, the message: "We screen on antique wooden separators and hand sort for optimum quality," signed, "from our family to yours."

Monika and Keith admit that the paperwork required to maintain organic certification is brutal. But they receive three to four times more for organic berries than they get for their wet-harvested, conventionally grown berries.

This is the first day the weather has permitted a dry harvest. It's sixty-one degrees Fahrenheit. The sky is blue with a few low, white clouds on the horizon. Keith, Gabrielle (foreman for the organic berries), Richard Turgen (son of the foreman who worked for Keith's father), and one other man walk slowly behind four machines about the size of small motorized lawn mowers, but with a handle on each side and escalating slats in the middle. As the men move in a pattern of ever-smaller segments, tines at the base of their picking machines pick up the berries and send them to the bottom slat. The slats then convey the berries up to the top of the escalating slats, causing them to fall into a burlap bag draped over the handles. When the bag is filled, the machine's operator stops, disengages the bag from the handles, and drops it on the bog.

On a larger, nonorganic bog, members of the crew would empty each bag into a larger bin. Then a helicopter would pick up the bins and deliver them to a waiting truck. On this bog the men do everything themselves. Richard is repairing machines when he isn't driving a picker-pruner in a parallel line to Keith. Gabrielle is supervising the packing of the berries for transport in between steering a picking machine.

I watch Tim Constantine, Keith Mann's other full-time worker, drive onto the bog in a red cart with a wagon hitched to it. The wagon seems oddly familiar until I realize that I have seen it in winter, fitted out for sanding, and in summer, delivering the pods for planting. Tim jumps down from the cart and grabs one end of the burlap bag. One of the seasonal workers takes the other. Together, they hoist the bag up to the level of the wagon, then let it fall. When the wagon is full of bagged berries, Tim drives it back to a farm truck. Two more men unload the bags for eventual transport to the screen house.

This particular harvest is all about simplicity. Each picker grabs a load of empty bags when he reaches one of the clumps of them on the ground. No bag folder has been hired to place a supply of bags on the handles of each picking machine. No helicopter zips down to lift bins of cranberries and empty them into a waiting truck. The full bags are hoisted by hand from the bog. There is a natural, almost primitive quality to these bogs and this operation. Poison ivy and other native plants flourish nearby.

Keith Mann's preferred picking machine, the one he is using today, is an older Weston machine. It sells for about a fifth of the price of a newer Furford picker-pruner, but its real advantage is that it has rounded tines, much like those formerly seen on antique wooden cranberry scoops, when twenty or so men, women, and children scooped berries while on their hands and knees. "The rounded tines cause less damage to the fruit," Keith points out.

Andrew Rinta in Carver thinks the dry-picking process "beats up the vines. And the yield is lower because a lot of berries are missed by the picking machines." Dry picking is limited to smaller growers, or the certified organic or European-certified larger growers. A midsize grower with the acreage of the Rintas can't justify doing it commercially. "It would take us six weeks, every day, to dry pick our bogs," Andrew explains. He used to dry pick on his grandmother's two-acre bog in nearby Rochester, but no more. "It was a pain. Each year, my brother and I would lug boxes over there. It was a real pain."

Many New England growers believe fresh fruit spoils faster if picked when wet. They don't begin picking until nine thirty in the morning, after the sun has dried the morning dew. They stop no later than six thirty, before the moist night air settles on the bog. They don't pick in the rain. They don't pick if frost is on the berries. Keith laments the lack of appreciation for the quality of berries he sells. He speaks of Whole Foods buying from a Canadian grower who wet picks and then dries the berries before shipment. "Then, only after that fruit has rotted, do I get a call from their buyer asking for my longer-lasting, dry-picked berries."

I pick a few berries to test their longevity at home. Left out on my kitchen counter along with some organic wet-picked berries from Oregon, the Mann dry-harvested berries were still firm three months later. After one month the wet-harvested berries were wrinkled and puckered.

On the Mann bog, dragonflies flit in the sunshine. Crickets chirp among the bins of cranberries in the truck. Red, yellow, and orange vines of the Virginia creeper weave their way through the evergreens. The silence of the woods envelops us. If I hadn't driven through a housing development to get here, I would think we were in a wilderness preserve.

These bogs were producing cranberries long before Keith's father, David, purchased them in 1976. Today, Keith expects they will harvest about seventy barrels per acre, or about seven thousand pounds of organic berries. Despite the extra labor, extra paperwork, and extra headache involved in growing, marketing, and distributing healthy, organic berries on these older bogs, the Mann family members believe producing a pesticide-free product justifies the effort. Ultimately, that way of thinking may go a long way in maintaining the cranberry's reputation as a healthful food, especially as more and more people choose to buy organic.

On the opposite side of the country, Ron and Mary Puhl have the advantage of wind, lots of it, to deter insects and reduce the need for pesticides. For over a hundred years, Mary Puhl's family has owned the property adjacent to Scott and Carol McKenzie's farm on Cape Blanco, Oregon. When she moved back to this hundred-acre farm, Mary and her husband, Ron, tried just about every occupation to enable them to live here. Ron bought a commercial fishing boat and fished for salmon. He dove for sea urchin. They ran six hundred head of sheep on the property. Mary worked as a bookkeeper for the nearby Port of Orford Port. In 1990, they decided to clear some of the gorse from a section of the property and build cranberry beds, as they are called in Oregon. At the same time, they built seven ponds for irrigation. By 1992 they had built five beds, covering over thirteen acres, and by 1993 they were able to begin planting. Six years later the farm was completed.

The year was 1999, just when the cranberry market crashed.

Christmas Day, Ron walked back to a cranberry bed, to see if any berries were still good enough to be used in a sauce for the family's holiday dinner. He noticed that the berries were deep red and plump and looked even better than they had in October. Ron bit into a berry and discovered that it was sweeter than the berries he had picked at harvest. After the holidays Ron and Mary took some of their late-season berries for testing. Testing results indicated a deeper, more even red color; a sweeter berry (9–10 degrees on an industry scale measuring sweetness versus 7.5 for berries harvested

earlier); and one with lower acid levels. From scientific studies cor-
relating color and health benefits, it follows that berries that stayed
on the bog longer had more anthocyanins and thus greater health
benefits.

So began the Cape Blanco production of premium, vine-ripened
cranberries that stay on the vine for approximately eighty days longer
than Wisconsin and East Coast berries. Mary and Ron planted 45
more acres and eventually purchased additional beds in nearby Bran-
don. Today, with 104 acres in production, they are Oregon's third or
fourth largest grower. Most of their berries are Stevens. Three newer
beds are planted with Grygleski, a Wisconsin hybrid.

The south coast of Oregon enjoys a unique combination of climate
and soil (terroir) that produces the sweeter Cape Blanco berry. The
temperature doesn't vary much from fifty-nine degrees in the summer to forty-two degrees in the winter. Berries come out of dormancy
in January or February, and flowers bloom in late April to early May,
four to six weeks ahead of Wisconsin and the East Coast. The threat
of frost is rare. Winds are clocked at upward of a hundred miles an
hour, but that doesn't have much effect on the ground-level cranberry, protected by a shield of leaves and vines.

Insects and weeds also have a longer growing season here. Winter
temperatures aren't cold enough to create ice on the bogs and thus
permit the type of sanding that encourages cranberry growth while
lessening weed growth. Mary's and Ron's vines are planted more
densely than those on eastern bogs or Wisconsin cranberry farms.
This method allows the vines to get established while leaving less
room for weeds. The Puhls' Cape Blanco beds are practically weed-
free. Ripe, dark-red berries can be seen beneath a lace mat of cranberry vines.

I ask Mary how the berries thrive in spotty sunlight and an environment so attractive to insects and weeds. "We don't have a lot of bugs because of the wind," she points out. "We work hard at trapping pests to determine the peak of their life, so that insecticides can be used when they are most effective."

Ron and Mary's cranberry beds make up a poster-quality picture of a well-tended cranberry farm. On one of their beds planted with the Grygleski hybrid, blocks of deep-red vines surrounded by six-foot-high woven wire deter deer, elk, cougar, and bear from eating or trampling the berries. I dig under the vines to pick a plump, red berry shaped like a small pumpkin (the defining shape of this Grygleski variety). As we cut it in half, we see the cranberry's four seed chambers. "It's the air in these chambers," Ron explains, "that allows the cranberry to float when a bog is flooded."

On the East Coast, water is more of a regulatory wetlands issue. On the Oregon coast, it's a matter of water rights. The coastal area south of Coos Bay, home to the bulk of the Oregon cranberry crop, has very little rain from the middle of June through September, causing growers to irrigate for two hours every other day during that period. New beds are designed to be terraced so that water runs back into self-contained water sources or channeled reservoirs.

When the Puhls filed for water rights through the Oregon Water Resources Department, all live-stream flow waters had been appropriated for other uses. From the first of May through the thirty-first of October, the use of stream water is prohibited. A grower can get winter water rights and save them, but it's necessary to specify what percentage of the water will be used for irrigation and what percentage for a variety of other uses.

For a few years Mary and Ron cleaned, sorted, dried, and sold other growers' berries. They built up distribution channels for the berries but ultimately decided to handle only the fruit from their own bogs. "Now we have complete control over each of our berries," Mary explains," and we can guarantee a customer that we have traceability to each of our bogs and the time that berry was picked."

At the Bandon bogs on the day of my visit, the Puhl family is harvesting in drizzle. Nick Puhl, Mary and Ron's nineteen-year-old son, is raking berries toward the conveyor belt that transports them up to the box of one of the farm's fifteen-thousand-pound trucks. Nick's grandfather is the driver. The Stevens berry being harvested is more elongated than the Grygleski, not quite as sweet, but of the same deep-red shade as the other late-harvested berries. When the truck is full, it will head north to a cleaning plant that uses well water treated with ozone, instead of chlorine, from its own self-contained wells. After cleaning, the berries will go to freezers to await orders from the stores and nutraceutical companies that make up the bulk of Ron and Mary's customers.

"We do extra." Mary says with pride. I ask her about the view on the East Coast that wet-harvested berries spoil sooner than berries that are dry harvested. "My berries should be used within three weeks," she answers, "unless they're frozen." They'll last for at least a year in the freezer. "For color retention and flavor," Mary advises, "cook with frozen berries."

About forty miles north of Bandon, the Coquilles have been growing and harvesting the Stevens variety on their reservation since 1995. Holding to the belief that land and humans exist to take care of each other, the Coquille tribal members grow and harvest only organic berries on the ten acres they call the Kilkich Farm. They dry harvest the organic berries.

The Coquilles are the only growers I met who publicly acknowledge greenhouse-gas emissions as a cause of climate change, as stated in a paper the tribe produced in 2010:

Regardless of the causes, members of the Coquille Indian Tribe are observing changes locally in the forest, prairie, streams, and ocean, in precipitation patterns, stronger storms, temperature extremes and fluctuation, and in the timing of seasons and processes like flowering and migration. Because of the potential risks climate change poses, the Coquille Indian Tribe is taking steps to prepare for the effects of climate change.

Bill Snyder, operations manager at the time I met him, told me that he and his family preferred the organic berry. "I think it tastes better," he says. "And it caters to a more local market." Customers of Whole Foods, Market of Choice, and Fred Meyer agree. In 2009 the tribe picked a hundred thousand pounds of organic berries, generating close to $250,000 in sales and making them the leading producer of organic cranberries on the West Coast.

In 2006 the Coquilles purchased two producing cranberry beds that together covered thirty-six acres. They weren't permitted to grow berries that qualified as certified organic on land that had been previously devoted to conventionally grown berries unless they dug up the beds, destroyed the vines, and replaced the soil. And they didn't want to destroy the land. They chose to wet harvest those beds and sell the berries as nonorganic, conventionally harvested cranberries.

When I visited, the Coquilles had just completed their organic harvest, but I was in time to observe the no-frills means they employ while harvesting their newer Seven Devils Farm on the old Seven Devils Road. The setting is spectacular. We drive parallel to the coast, going steadily up through a forest of spruce, hemlock, cedar, and pine. At the entrance, a Cooper's hawk perches on the gatepost, guarding its territory. Emerging from the forest, I first see an expanse of blue sky meeting the deep red of cranberries. In the distance, rollers can be seen flowing over the Pacific and crashing on the rocks below. Arctic terns flit over the beds much as swallows do at the Piney Woods harvest.

The Coquilles employ four full-time employees, two seasonal employees in the summer, and three or four seasonal workers at harvest. After harvesting, the berries are taken to a blueberry plant, where they are thoroughly cleaned, sorted, and packaged. All the Coquille berries that are wet harvested are processed. Bill estimates that 70 percent of them are turned into either juice or juice concentrate to be privately labeled. A substantial portion of the remaining 30 percent become individually quick-frozen (IQF) berries, a product peculiar to the West Coast. Most of the remaining berries are dehydrated and infused to become sweetened dried berries for cereal, trail mixes, granola, and baking. A small percentage is turned into powder for the relatively small cosmetic and pharmaceutical markets, where they can be found in face powders, capsules sold to ward off urinary tract infections, and sachets sold as probiotic supplements.

This is wet harvesting stripped to its basic components. Farm manager Neil Mong, Duane Long, and Santiago, the husband of a tribal member, are the only people working the harvest. Neil is on his cell phone, making the arrangements for cleaning, sorting, and processing. Duane and Santiago alone are harvesting the berries. Strong and wiry as a bronze Giacometti sculpture, Duane is in the far corner, surrounded by a wash of deep red. Up on this plateau, the wind is steady at twenty to twenty-five knots. It blows the berries across the water to where Santiago corrals them on the lee side.

Somehow, Duane, always good humored, is both in the water and out of the water at the same time. He sets up the escalator and balances a straight ladder against the side of the truck. He drags a water hose to the edge of the dike for runoff. Then he jumps back in the water with the rake, runs up the ladder to check the level of the berries in the truck, rakes the berries toward the back of the truck, periodically jumps in the driver's seat to move the truck forward as needed, then back in the water to rake the berries toward the escalator. He yanks off one layer of clothing when wet with perspiration but still manages in the wind and fifty-degree temperature not to break stride. When the white Coquille truck is full to its fifteen-thousand-pound capacity, he drives it and the cranberries to the handler.

Later, in Bill Snyder's office, the call comes in from Duane reporting that he delivered two truckloads totaling thirty-five thousand pounds of berries from bed number 8.

These berries are the largest and most uniformly red berries I have seen harvested from any cranberry bed, marsh, or bog. They were harvested in one afternoon with no suction device to deliver the berries to the conveyer, no scaffolding to permit viewing of the amount of berries in the truck, no designated truck driver, and only two men. It is by far the most efficient cranberry harvesting operation I have witnessed.

In 2012 the tribe converted the Kilkich Farm from an organic farm to a conventional one. Todd Tripp, the farm's present manager and director of planning for the Coquille Tribe, attributes the decision to the arrival of the lotus weed, an invasive legume that takes over a cranberry farm once the weed has become established. Tribal members tried for several years to eradicate the weed from the bogs by pulling each lotus plant out by hand, but the weed was winning the battle. The tribe finally decided to use chemical herbicides as opposed to giving up growing cranberries altogether. Todd foresees the possibility of returning the Kilkich Farm to an organic farm in three years time if he is able to control the invasive plant.

Dot and Jack Angley's Flax Pond Bogs in Carver, Massachusetts, are a sharp contrast to the seclusion of the Coquille farm. At a triangle in the dirt road, a painted sign proudly proclaims these bogs to be part of the Ocean Spray Cooperative. Farther along, "Visitor Parking" signs mark a dirt lot carved out for cars and tour buses. The Flax Pond Bogs clearly welcome and cater to tourists. A timetable for touring the bogs is posted at the door. In the parking lot, near portable toilets, a motor coach awaits the return of its passengers from a guided tour of the bog.

Cranberry vine wreaths invite the visitor beyond the open doors of a lovingly restored screen house, where Dot Angley, her granddaughter-in-law, and three-year-old great-grandson all greet the visitor, each in his or her own way. Dot answers questions on cranberry lore, history, growing, and harvesting. Her granddaughter-in-law weighs fresh berries and tallies up cranberry items to be purchased. Aside from a wooden Bailey separator machine for sorting berries, the interior resembles a country store from the turn of the twentieth century. Available for purchase are cranberry mugs; cranberry candles; chocolate-covered cranberries; the Angley family cranberry bog honey; jelly made from cranberries, Vidalia onions, and peppers; and Bog Wash S-lime-y foaming liquid soap with a rubber frog in a bottle.

Outside, cranberry bogs stretch in three directions. One of them, the closest to the screen house, has been in operation as a commercial bog since the 1890s. Local lore has it that flax was commercially grown on the property before cranberries. Flax continues to grow around the edges of the bogs, giving credence to the legend and authenticity to the bog's name.

The Angleys grow Early Blacks, which they harvest and sell as whole fresh fruit. All their bogs are dry picked. Asked about the loss of yield and stress to the bog from dry picking, the unflappable Dot unequivocally states, "only if you have inexperienced help. All our crews are experienced pickers."

In the distance eight men are working on one bog, and two on an adjacent one. The temperature has risen to sixty-seven degrees. Puffy clouds dot an otherwise blue expanse of sky. Warm sun has caused one man to strip to his waist, revealing tattooed arms. The men either push picking machines, empty bags of cranberries into stacked bins, or load empty burlap bags onto the arm of the picking machines. Conversely, only five men wet harvested the Piney Woods bog of similar size and age. The yield was higher on that bog than what is expected from this bog. But Ocean Spray pays more for dry-picked, whole berries. "And," Dot says while echoing Keith Mann, "the whole dry berry lasts a lot longer."

Dick Nantais, a retired engineer, has been picking at Flax Pond for eight years. He slowly walks behind the picking machine, letting it move across the bog at its own speed as he replaces full bags with empty ones. A rhythmic squeal in his machine becomes louder until finally the machine stops, causing Jack Angley to arrive with an oil can. "These machines have a lot of moving parts to break down," Jack comments as he stoops to make the necessary repairs. I ask him about the lone man who has been steadily operating a picking machine, always one bog ahead of the others, the man whom Jack and Dot's grandson Matt describes as "my grandfather's right-hand man."

"Norberto has been with me for sixteen years," Jack tells me. "He and I run forty-two acres here. Plus six in the woods for another person." Norberto is the only full-time employee at Flax Pond. All the others, many of them relatives, help out as needed. Some, like Dick Nantais, are seasonal.

Until recently, independent growers were paid more for berries than Ocean Spray paid its growers. The past few years have seen the reverse. In 2012 independent growers received approximately $22.00 to $28.00 per barrel; the 650 Ocean Spray growers in the United States received $63.11 per barrel for A Pool (top color) cranberries. Much of that difference accrues from the 90 percent of Ocean Spray products that are "value added," such as juice blends. To be included in the cooperative, member growers must guarantee that Ocean Spray receives 100 percent of the berries they harvest. Ocean Spray has maps of each member grower's bogs, and the cooperative takes aerial photos for verification and to monitor individual holdings. As a result, it has a record of all the acreage devoted to growing cranberries in the United States. Growers, like the Angleys, are allowed to keep some acreage for growing fresh fruit, but only if Ocean Spray agrees to exempt that acreage from the grower's processing agreement. The cooperative allows Dot and Jack to withhold two thousand pounds to sell in their store. Because Jack and Norberto pick for a few other growers, the Angleys are able to use the additional allotments to supplement their supply for the store as tourist demand increases.

I walk up the hill and look back at the bogs, where gray bins of cranberries are stacked, strapped, and waiting for a helicopter to pick them up. At the edge of the bog, a Flax Pond truck is parked and ready to transfer the berries to Ocean Spray's receiving station. Next to the shop, one of five original bog houses has been restored to its original condition, when a bog worker's family would have lived in the one-room "up/down" building. Inside the fourteen-by-eighteen-foot structure, stairs lead from one corner to the sleeping loft overhead. A table, sink, icebox, and a couple of chairs, along with a wood-burning stove for cooking, hot water, and heat, fill the remaining space. The surrounding porch allowed workers a view over the bogs to the woods plus the chance to enjoy the summer breeze. But from the time the mist left the bogs until sunset, they and their children worked on their hands and knees without the benefit of machines. Much has changed, but the beauty of a cranberry bog remains.

Less than twenty-five miles offshore, Tom Larrabee Sr. recently celebrated fifty years as foreman for 291 acres of cranberry bogs on Nantucket Island, amid some of the most valuable and desirable land in the United States. Some of these vines have been cultivated for more than a hundred years. They provided the cranberries carried on Nantucket whaling ships to ward off scurvy. Today they are owned by the Nantucket Conservation Foundation, created to preserve and protect the bogs and the land surrounding them.

Situated in the middle of three thousand acres of conservation land, the Nantucket bogs have no abutters and no one to complain about the sound of helicopters or picking machines during harvest. Thirty-five acres are organic. The bogs are located in an area of isolated summer homes whose owners leave the island before harvest. One home's owner near the organic bogs is on the board for the foundation and recently signed a waiver to allow drainage over his property. He, like many of the people who bought or built homes on the island, is grateful to the foundation and the cranberry bogs for helping to preserve the land and a way of life he and his family originally came to Nantucket to find.

This part of the island is a lush plain exposed to winds and fog coming off the Atlantic Ocean. Nicknamed the "Serengeti," it is one of the last remaining pristine grasslands left on the East Coast. One hundred years ago much of the island was covered in native grasses until sheep grazing ceased and scrub pine began to take over. "If it wasn't for the Conservation Foundation," Tom says, "this place would be a golf course."

"Or house lots," his son adds. The present manager of the Nantucket bogs, Tom Jr. is as deeply committed to maintaining the integrity and sustainability of this cranberry farm as his father. "This cranberry bog," he says with evident pride, "it will always be here. And it will always be producing a crop . . . as long as the foundation board reaches into its pockets on the bad years." He adds with confidence, "And they will."

Almost 50 percent of Nantucket is preserved land. One third of that, about nine thousand acres, is owned by the Nantucket Conservation Foundation. Board members tour the cranberry bogs two or three times per year and ask what they can do collectively to help the Larrabees grow cranberries. "They have a genuine interest," Tom says. "It's written in the bylaws that these cranberry bogs will always produce a crop. And that's what saved us because this could have easily been left to become wilderness. And it takes a lot longer to bring it back than to let it go."

Because these bogs are owned by a nonprofit, the Larrabees are somewhat sheltered from the vagaries of the market—in 2012 the price per barrel received came in at only about twenty dollars a barrel—but they have an incentive to produce the highest yields while looking to the long-term health of the vines. "My father kept some of these vines alive for over fifty years," Tom Jr. says. He laughs as he adds, "Prying information out of him is a constant battle, but he knows so much about growing cranberries out here."

"Well, being here for fifty years helps," the senior Larrabee interjects.

The practice of growing cranberries on the mainland varies little from the process on Nantucket, except that the latter is an island. The Larrabees don't have the luxury enjoyed by mainland growers of sharing or leasing equipment from other growers. Here there are none. Everything they need, they own.

After harvest ends in November, the bogs are flooded to protect the vines from the cold and wind. "Come January, if it freezes hard enough, we can put down an inch of nice, screened sand," Tom Jr. explains. "Ideally, every three years." In the absence of ice, they drive on the bogs with the same machine used to loosen the berries at harvest, but with soft wheels to put the least stress on the vines.

Tom Sr. describes the effect in March when the ice melts as "a positive charge meeting a negative ion," releasing the nutrients that are locked up in the plant during its dormant period. By the end of April, on Nantucket, the skies are typically clear and the wind has abated. Then, the cold air sinks to the level of the vines just when the bud is at a fragile stage and can only tolerate a thirty-degree temperature.

The present irrigation system was designed about thirty years ago. That's about to change. The foundation just had a grant approved to install pop-ups and a new system. The grant doesn't cover labor, but as Tom Jr. says, "We do all the labor ourselves, so it will all work out."

Two gating factors are unique to the success of the Nantucket crop. One is the Massachusetts Steamship Authority, which controls the Martha's Vineyard and Nantucket ferries. The other is the weather's impact on the ferry schedule. "June 19, I've got 450 beehives coming across with enough honeybees to pollinate the flowers," Tom Jr. explains. Tom made the ferry reservations for those bees a year in advance, including specific reservations for the return trip. Arrival of the bees can take place only after the bees have finished pollinating the Maine blueberry crop. There's little wiggle room.

On the morning of October 14 I manage to get a place on the ferry for the second and last day of harvest on the foundation's Windswept bog. The temperature on the island is thirty-three degrees, and the wind has been gusting to thirty-five knots. When I arrive at the bog, I realize how it got its name. Though organic, this is a wet harvest, and when the berries leave the ferry on the mainland, they will be trucked to the Decas cranberry company for processing, then sold as organic, sweetened dried berries.

Decas buys everything the Larrabees can produce—both organic and nonorganic. But getting the berries there isn't simple. Transportation to and from the Decas plant last year totaled $18,000. "We have three trailer trucks going and three trucks coming back on the ferries every day until the harvest season is over," Tom Jr. tells me. When he reserved ferry space, he didn't know how many berries he would be transporting or exactly what dates the berries would be ripe for picking. Most years he books space from the middle of September through the middle of November to allow for some leeway for a bumper crop or a spate of bad weather.

Cranberries have always afforded a gateway to America. These bogs are no different. Hristo, the Windswept foreman and only other full-time employee, is from Bulgaria. He speaks Bulgarian, Russian, Spanish, and English. Now an American citizen, he lives on the island with his wife and two children. Two part-time workers on the bog are from El Salvador and leave after harvest to return to their larger island.

The wind has blown the berries to the far end of the bog being harvested, and the men are in the water in their black waders, slowly tightening the boom to corral the berries back toward the elevator and the waiting truck. Tom has confirmed space on the ferry for the truck to return to the mainland, but he can't afford to send a half-empty truck, and Decas can accept berries only until five. The company will keep employees working until seven if the Nantucket truck can make the scheduled ferry departure. It's a three-and-a-half-hour ferry ride, and only one lane on the Sagamore Bridge is open due to bridge repairs over the Cape Cod Canal. One water reel has broken down, and Billy, the full-time bog mechanic, has yet to arrive with the repaired part. It is not looking good.

Tom Sr. explains that the older water reels have rotary reels; the newer machines have fork-like springs. "This is a thirty-year-old water reel, a patented Wisconsin machine." He shakes his fingers and looks at the sky, indicating disdain.

These organic bogs were planted in the 1920s. One of them is still planted with an heirloom called Black Veil. That bog gets a little more shade, making it particularly sensitive to fungus, and one of the characteristics of Black Veil is its ability to resist mold and other pathogens. The yield is 70 barrels per acre from these bogs, as opposed to the 150 to 250 yield produced by his conventionally harvested bogs planted with Stevens and Ben Lears. But Decas pays more for the organic berries, and the ideals of the foundation are more in tune with organic farming than with nonorganic.

When the cranberry market crashed, at the end of the twentieth century, Northland, the consortium that employed Tom to manage the bogs, broke its lease with the foundation. From 1998 to 2001 Tom Sr. single-handedly kept these organic bogs producing. He was on his own, but he wasn't going to abandon "his bogs."

Today he is operating one of the water reels. Before it broke, Richard Mack, age seventy-seven, was operating the other one. For more than twenty years, the two older men have been driving the water reels in tandem, guided by the orange and yellow flags marking the sections of the bog and any underwater obstructions. Now both of them are bent over the broken machine discussing with Billy how best to repair it. Until now, the harvest was moving seamlessly from bog to bog with the two water reels always one bog ahead of the corral process.

"I have one bog left, number 9," Tom Jr. says with dismay, "and that's not even close to being flooded. I need both machines in operation."

A cranberry grower needs to be a welder, a mechanic, an engineer, a soils manager, a carpenter, a mason, an electrician, and a farmer. He or she has to love being outside in all kinds of weather and to have the patience and confidence it takes to survive and protect the crop. As the men kibitz over the best way to repair the water reel, swallows swoop and dive for insects. Tom Jr.'s black lab, Emma, and his father's spaniel, May, jump in the water to catch the unsuspecting vole swimming in search of a new home to replace a recently flooded one. Hikers and dog walkers occasionally pass through the area, and both Larrabees stop their deliberations to greet them, exchange island pleasantries, and, for the uninitiated, explain the workings of a cranberry bog.

When I ask the senior Larrabee what has kept him on an island farming cranberries in all kinds of weather, he answers, "It's quite a challenge to take the bogs from January to harvest and end up with a successful crop, but everyday I look out my window and tell myself how lucky I am to be here."

Tom Jr. recently built a bog manager's residence for him, his family, and future managers on upland overlooking the bogs. From the house, you can see the Sankaty Head Lighthouse in the distance. "I joined the Marine Corps and left home for twenty years," he says, "but being here . . . it's hard to put words to it. I mean it looks the same now as when I was little growing up here . . . mowing grass out here."

"And it's not going to change," his father adds. "There are no houses you can see through them hills. The foundation owns it all." He sweeps his weathered hand toward the view. "And we are lucky enough to live and work here."

10

PROCESSING, HANDLING, AND DISTRIBUTION

The cranberry business is no longer looked upon as speculative. It now takes its rank among those legitimate occupations which make good returns for well-bestowed labor; but like any other business, to be pursued profitably, it must be conducted upon right principles, and with strict attention to details.

—Joseph J. White, *Cranberry Culture*, 1885

Despite an offer to stay late by Parker Mauck (director of Grower Relations for Decas Cranberries), despite the Larrabees' best efforts to keep the beaters working, and despite the truck driver's agility, the Nantucket berries did not get processed the day they were picked. Instead, they were on the next morning's ferry to the mainland. Five hours later they were unloaded at the Decas processing plant, cleaned, sent through a drying system, and placed in bins heading to cold storage—all before noon.

Decas Cranberry Products in Carver, Massachusetts, is a story of hard work and endurance and a family that has always believed it can accomplish whatever it sets out to do. In other words, it is the American immigrant story. It is also the cranberry story.

John Decas, a gentleman in boat shoes, denim shirt, and cloth cap, gave me a tour of his plant. Along the way, he shared some of his family's history. His father, age thirteen, came to Boston from Greece in 1905. An uncle met him at the dock, took him to New Bedford and put him to work. The first year he made ninety-six dollars shining shoes. He sent half of it home to his family, who, as John describes it, "were living in dire poverty, with ten kids, up on a mountain where there wasn't much you could do to earn a living. They'd lived through the Turkish oppression and then World War I."

After two years in this country, John's father was joined by an older brother, who had previously sailed to New York, had been held at Ellis Island for two weeks, then sent back to Greece because no one had been at the dock to meet him. Later, when he returned to the United States, the two brothers bought a one-eyed horse and cart. They loaded the cart with fresh vegetables purchased from nearby farms then headed up the coast to the wealthier towns of Mattapoisett and Marion to sell their goods. In Fairhaven, Franklin Delano Roosevelt's mother and her summer friends took pity on the boys and instructed their kitchen staffs to give them breakfast or lunch, depending on when they arrived. In addition to putting meat on their bones, familiarity with the kitchen staff provided the brothers with the opportunity to sell their produce.

Word spread, as did the number of customers, and by the 1930s the business was thriving. The community had gotten to know them. A third brother arrived, and with help from a local benefactor, the three Decas boys were able to move to nearby Wareham, where they rented a warehouse and expanded their territory to include Cape Cod. Over time they had saved enough to buy a block of stores in downtown Wareham and open a grocery store. One morning in 1934 the brothers, now relatively seasoned businessmen, were loading up their produce in the warehouse. Leck Handy, a neighbor and popular cranberry grower who sat on the board of the local bank, opened the warehouse door and told them about a ten-acre cranberry bog in nearby Rochester that his bank was about to foreclose on. And that was the beginning of Decas Cranberry Products.

Today, the Decas family owns 450 acres of bogs and a sixty-thousand-square-foot plant. They are the largest independent handler and processor of cranberries and, until 2012, when they sold that component, they were the leading provider of cranberry ingredients for nutraceuticals. To independent growers John Decas is a well-respected fellow grower and a champion of their causes. To Ocean Spray, the growers' cooperative, he is a force that won't go away.

Cranberry cooperatives had existed since 1864 for the main purpose of protecting growers from unscrupulous dealers. The following handwritten letter from a New York dealer to one of his cranberry suppliers gives a sense of the general distrust on both sides:

WILLIAM CROWELL & Co.
WHOLESALE DEALERS IN CAPE COD CRANBERRIES
254 GREENWICH STREET, NEW YORK

October 26, 1881

John H. Baker
Dear Sir,

We sold all your boxes today 2 ## 50—one X 337; one X 3¼ also sold your 15 barrels @ $10. If you will send along your berries we will get the highest price. I hope you will send some of your berries to the gentleman you write about, he is trying in every dishonest way to hurt me.

Yours truly,
W. Crowe

In 1888 the Cape Cod Cranberry Growers' Association was formed. Its purpose was to standardize berry color and size, market cranberries from bogs in Massachusetts, and provide a vehicle for the exchange of ideas on cranberry cultivation. Seventeen years later, in 1905, Arthur Chaney, a wholesale fruit dealer in Iowa, teamed up with two other buyers to purchase that year's Wisconsin cranberry market. The following year was a great year for growing cranberries. The next year found the industry with a surplus, an early example of what would become cranberry growers' typical boom-and-bust pattern. In an effort to return market stability to where it had been and to attempt to protect themselves from similar patterns in the future, growers in the Midwest decided to join Chaney to form a cooperative of midwestern growers—the Wisconsin Cranberry Sales Company.

At the same time, around 1907, the New England Cranberry Sales Company, under the label "Eatmor," established standards of quality and handled the marketing and distribution of its growers' berries. Chaney convinced members of the New England Cranberry Sales Company and the New Jersey Cranberry Sales Company to join him in creating the American Cranberry Exchange.

The Exchange was a marketing cooperative whose member growers elected their governing board from their own ranks. Members delivered their entire crops to their local sales companies, and the Exchange provided marketing, sales, and transportation of the berries to the customers. It also helped stabilize prices for the growers. Member growers were paid a partial payment upon delivery of their berries. Then after the year's harvest was completed and sales tallied, they were paid a portion of the total sales weighted for the quantity and quality of berries delivered. Cranberries were sold only as fresh fruit, and quality was taken seriously by both growers and the public. Berries not up to a certain standard of color and uniformity were dumped, thus diluting growers' profits. Then in 1912, Marcus

Urann, a Maine lawyer who had begun to buy cranberry bogs six years earlier, saw a way to use the white berries along with the red. Canned products were replacing home bottling and pickling for the American housewife, and Urann began to can mixtures of both his white and red berries as sauce in Hanson, Massachusetts, under the Ocean Spray label.

From its inception the executives at Ocean Spray have been businessmen, not farmers, although the present TV ads featuring actors as farmers obscures that fact. When Marcus Urann coined the name "Ocean Spray" as a label for his canned cranberry sauce, he was a recent graduate of the Boston University Law School and a practicing lawyer. One of his clients was a cranberry grower who let his lawyer in on the little-known fact that cranberries offer a high return on investment to the grower who knows how to grow them. Urann initially bought a small bog, followed by more land acquisitions. He soon recognized that if he could find a way to sell the lower-grade berries he could also show a greater profit. He began buying other growers' blemished berries at reduced prices to put into his cans of cranberry sauce.

Everyone benefited, at least initially. The grower who used to discard berries that weren't perfect now had a market for them; the canner could pay less than the market price for top-grade berries and was able to make a greater profit from the lower investment in materials. The problem was that at least two other growers were also canning berries: John Makepeace, son of Abel (A.D.) and at various times the grower with the largest acreage in Massachusetts, and Elizabeth Lee, with similar holdings in New Jersey. Processed cranberries in the form of sauce and juice were outpacing the sales of fresh berries by about 20 percent each year. Competition between the three canners was further depressing the market for the top-quality berries that sold as fresh fruit and provided the bulk of a grower's annual income. In 1930 Urann suggested a merger between the three canning companies. Only after he had drawn up the papers did he check the legality.

This was during the Depression, and antitrust laws prohibited formation of companies where the new entity would control more than 90 percent of the market. Urann turned to the Capper-Volstead Act, which exempted an agricultural cooperative from antitrust laws. He claimed that the three canners complied with the act because they were primarily farmers, not canners. In return for compliance, the three grower-canners agreed to turn over their canning operations to the newly formed growers' cooperative. Marcus Urann by that time owned the largest canning operation of the three. Despite the fact that John Makepeace was the world's largest cranberry grower, with holdings of about 1,500 acres of bogs, he reluctantly agreed to Urann's becoming the largest stockholder and president of what would ultimately be called Ocean Spray.

Booz Allen Hamilton, in a report on the cranberry industry, later reported that although "his intelligence and business capacity is almost universally admitted to by both his friends and his foes . . . nevertheless, it is an outstanding fact, as revealed by survey interviews, that more than two-thirds of the growers do not fully trust Mr. Urann or his motives." John Makepeace was of the majority opinion, but realized the wisdom of doing business with Urann.

Before the formation of Ocean Spray, Marcus Urann recognized and capitalized on the profitability of cranberries. In a 1909 prospectus for the United Cape Cod Cranberry Company, where he was one of the officers, a photograph shows what is listed as "Property No. 3," with the caption "a part of which paid 100 per cent profit in the third year." It could be a prospectus for any commodity. It just happened to be cranberries. Today board members at Ocean Spray are, for the most part, cranberry growers. Independent growers who are not Ocean Spray members find that, as John Decas points out when speaking of the cooperative, "We share cranberry issues of mutual interest, but as independents, we try to get together to protect ourselves from something Ocean Spray may want to impose on the entire industry that may be good for Ocean Spray but not good for us."

For both Decas and Ocean Spray, China is a huge potential market. The Decas company is presently working with a distributor of sweetened dried cranberries to bakeries in China. John Decas fears that it's inevitable that the Chinese will eventually learn to grow their own cranberries. "They have badly damaged the apple business in this country when they decided to grow apple trees." he says with a touch of envy. "They've got so much land, so much cheap labor, and so little in the way of environmental laws to contend with."

The 1934 contract the Decas brothers had signed to purchase their first bog included a stipulation that they join the New England Cranberry Sales Company. But, they wanted to sell cranberries along with their produce, not through someone else's system of distribution. Shortly after the bog purchase, the truckers' union staged a strike at the Atlantic and Pacific Stores (A&P), then the largest national chain of grocery stores in New England. Despite tire slashing and gunshots, the brothers used their trucks to deliver produce to A&P until the strike was over. The risk was offset by the gratitude of A&P management, who entered into an agreement with the Decas brothers to carry their fresh cranberries.

Instead of renewing a contract they didn't want with what was by then called Ocean Spray, they took a look at the sales of Ocean Spray's canned sauce and quickly determined that cranberry sauce might allow them to also use some of the berries they couldn't sell as fresh fruit. "When you sell fresh cranberries," John Decas explains, "there's always a certain percentage of the berries you acquire that don't qualify to be sold as fresh fruit. They might be a little too green;

they might be unevenly colored; they might be too small." The brothers contacted another Greek transplant, the owner of a vegetable-canning company in New Jersey. The Decas men must have had a compelling story because the company's owner agreed to try canning their cranberry sauce for that Thanksgiving season. It was a wise decision and the Decas canned cranberry-sauce business took off.

Until the early 1940s fresh cranberries were sold from staved, wooden barrels located near the cash register at the local country store. Each time a customer wanted to buy berries, the grocer would scoop out the desired amount, estimate the weight, and empty the scoop into the customer's shopping bag. A&P decided to promote preweighed, premeasured, prepriced one-pound packages, wrapped in cellophane and ready to be paid for without the back-and-forth scooping, weighing, and folksy conversation in between. This may have been the first bell in the death knell of the country store. It was also, John Decas tells me, "the first prepackaging of cranberries."

Success created a new problem: supply, a problem as familiar today as then. The Decas brothers didn't have enough cranberries to fill the demand. They began contacting other growers, saying, "bring us your berries." Little by little, fruit began to flow in from other growers, and Decas became handlers, competing with Ocean Spray.

"Ocean Spray's a great company. But bigger's not always better," John insists. "They have a huge bureaucracy." It's important to remember that Ocean Spray is a publicly held company. Growers have to purchase Ocean Spray stock to become members. The company doesn't offer stock options to attract management. They pay bonuses. And in the volatile world of cranberry supply and demand from one year to the next, that money has to come from sale of the growers' harvests in both good years and bad years.

"So now we come along," he says. "We were a small handler who only sold fresh cranberries. What we were best known for was the packaging of our fresh fruit for the holiday season, sold under our Paradise Meadow brand."

Today Decas Cranberry Products is the largest cranberry-ingredient company, handling, storing, and transforming the berries from their own bogs and those of other growers around the United States and Canada and selling to more than twenty countries. A large part of the company's product goes into making extracts used in food supplements, as either liquids, capsules, powders, or gels. The competition is the ingredients division of Ocean Spray. Since 2011 Decas Cranberry Products has focused on adding value to the Paradise Meadows line with individually packaged products aimed at a health-conscious market. OmegaCrans is a line of sweetened dried cranberries fortified with omega-3 oils derived from seeds of the berries; LeanCrans have 50 percent less sugar than standard dried berries; and the Funny Face brand is made with less sugar, added fiber, and the enticing names and flavors of Goofy Grape, Rootin' Tootin' Raspberry, Freckle Face Strawberry, and Choo Choo Cherry, all aimed at the children's market or the market for parents with children. In 2012 the company contracted with a company in Mexico that will distribute an unbranded version of Funny Face throughout South America.

When a truckload of cranberries arrives at the Decas processing plant, no probes penetrate the truck's contents to take samples. A notice in the Decas waiting room shows growers where to leave color samples. By comparing samples left by other growers, a grower can gauge when to harvest to get the best price. Both Ocean Spray and Decas pay growers an incentive for high "TACY," or good color. "For the grower, it's a tough decision," John adds. "He can't wait forever. Winter's coming. Maybe it's a family operation and the family members can harvest only on weekends. They've got to get the cranberries in. The longer you wait, the higher the risk. On the other hand, the longer you wait, the higher the color; the higher the color, the higher the incentive."

Consolidation in the food industry and supply drives much of the market for cranberries. As companies acquire other companies, their buying power increases. Walmart and Costco buy in huge quantities, giving them the bargaining power to depress prices to undercut the smaller competition. Wal-Mart's organic and natural foods division is promoting the small farmers. It's too early to see how that plays out, but independent farmers who depend on Walmart for distribution of their whole crop could be scurrying for a market for their berries should Walmart change its policy.

Pappas and Cliffstar, two other dominant private-label handlers, also buy cranberries from various growers and turn them into "a little bit of sauce . . . mostly juice," John says. "And they pack it under whatever brand . . . Stop & Shop, Walmart, you know that kind of thing. And they have these huge contracts with these big chains and superstores that want products under their own brand. They don't sell ingredients."

"I was the farmer in the family," John says. "I managed the farms and then built the packing plant. It was obvious to me, we could no longer be just a handler of cranberries doing the same old thing." He took a look at how big Cliffstar and Pappas were becoming and realized that Decas couldn't supply them with all the berries they needed. He didn't want Decas to be the middleman competing with his customers for his own growers. So what next?

Because cranberries are difficult to grow and largely unprofit-
able in their natural state, cranberry farmers must be resourceful
and ingenious. Thus cranberry farming is the very essence of Yan-
kee ingenuity.

—Angus Kress Gillespie, "Cranberries,"
1999

11

HYBRIDS TAKEN TO A NEW LEVEL

At first the cranberry yielded to the growers' large profits. This led to increased production. The supply has at last overtaken the demand.

—Judge John A. Gaynor, *Milwaukee Sentinel*, 1900

What would ultimately be a recurrent plague first appeared at the turn of the twentieth century. From that time on, periodic oversupply became more destructive to growers' profitably than some of the most troublesome insects. From 1900 to today the fortunes of cranberry growers have risen and fallen regularly as a result. The year 1899 produced the highest cranberry crops since Henry Hall's first shipment of cranberries to the New York market—190,000 barrels in New England; 70,000 barrels from New Jersey (that number was later scaled down due to a late frost); and 35,000 barrels from Wisconsin. Conversely, the expected crops for 2012 were 2,100,000 barrels in New England; 542,500 in New Jersey; 4,500,000 from Wisconsin; and 400,000 barrels from Oregon, which wasn't even being counted in 1899.*

---

*A barrel equals one hundred pounds of cranberries.

Owing to the fact that cranberries keep longer than other fruit, orders can be placed as needed, and the fruit once harvested is stored by the grower. If orders aren't forthcoming, the berries can be used to fill next year's orders, but due to the plant's sixteen-month cycle, the next year's crop is already growing. Cranberries are sold as a commodity, and the laws of supply and demand rule the market. The price paid for the 1900 crop was equal to or below the cost of production for most growers.

What to do? The most obvious answer is and was to increase demand. Since 1900 organizations have been looking for ways to expand markets for the berry. In 1932, to help motivate buyers, then congressman from New York, Fiorello La Guardia, issued a collection of cranberry recipes, and in 1949 Ocean Spray circulated the following ad in an effort to get people to buy cranberries throughout the year, not just at Thanksgiving:

What's Chicken
without
Cranberry Sauce!

What Sterling is to silver
What honey is to a bee
What donuts are to coffee
CRANBERRY is to me!

Fifty years later, following another surplus and resultant crash, independent growers' organizations looked to foreign markets to absorb future crops. Today close to 30 percent of the harvest, about $133 million, is exported. Around 5 percent of Ocean Spray's exports head to Asia, the overseas market with the greatest potential for expansion and absorption of the berry's greater yields. Australia, Great Britain,

Canada, and Germany make up the largest buyers of cranberries outside the United States. For years French women have been buying cranberry juice to ward off urinary tract infections but, until recently, only with a prescription from their doctor.

Despite the cultivation of overseas markets, the expansion of cranberry growth to Canada, Chili, and other countries, together with the increased yields of the new hybrids, continues the cycle of conflict between oversupply and demand. In 1996 a barrel of fresh whole cranberries sold at sixty-five dollars, but by 2001 the price had fallen to eighteen dollars per barrel. In 2013 a barrel of cranberries sold for between twenty to thirty dollars, barely covering the cost to grow them.*

Enter the hybrids, and the oversupply picture becomes a classic tragedy of the commons in reverse. Each grower wants to see his bog, marsh, or farm produce a higher yield than in the previous year, but the optimum conditions for the community as a whole depend on limiting the yield to fit the demand. In most crops 50 percent of increased productivity is due to selective breeding and hybridization or gene shuffling. And the desire to find a disease-resistant berry with a higher yield is in the blood of every cranberry grower.

John Decas still grows cranberries on a bog built in the 1880s. He echoes Nodji Van Wychen when he says, "The older berries produce the very best fresh fruit. They're smaller, hardier, prettier . . . and they look good in fresh fruit packaging." Larger berries are more fragile and more prone to denting and cracking. Building new bogs, converting to new hybrids, and waiting out the years before the new vines produce sizable crops is a long, arduous process. Stevens, which

---

*Figures listed for sale prices of a barrel of cranberries are the price an independent grower would receive. Ocean Spray growers have recently been paid twice as much or more per barrel.

are now being replaced by even higher-yield hybrids from Ed Grygleski, Abbott Lee, and others, were introduced in the 1950s and took another fifty years to become as prevalent as they are today.

Ed Grygleski Sr. knows he is a lucky man, living the life he chose, on the spot where he grew up outside Tomah, Wisconsin. The youngest of his four sons and his daughter live in homes they built next door. Six grandchildren play where he played as a boy.

Cranberries have been harvested on this 2,600 acre property since 1891. At that time a LaCrosse consortium owned the land as an investment and paid local farmers to handle the growing. They called their venture the Valley Corporation. Ed's father managed the property for the owners in 1939, then was able to purchase it from them in the 1950s. Ed took over management after his father's retirement, and in the 1970s he began to experiment with breeding new cranberry varieties. Today he is considered a pioneer grower-breeder, and the Valley Corporation is providing Grygleski hybrids to Mary and Ron Puhl, plus other cranberry growers in Wisconsin, Massachusetts, New Jersey, Washington, and the Canadian provinces. A grower who places an order for Grygleski hybrids will be buying cut vines, not pods.

Ed Sr. and his son, Ed Jr., and I are in one of the family's pickup trucks, driving through their marsh (the Wisconsin equivalent of a bog), when a pair of sandhill cranes gracefully lifts off from a nearby cluster of reeds. "We have twelve to twenty-four nesting on the property," the senior Grygleski comments. "Maybe we'll see the bald eagles or osprey while you're here. We have only one pair of loons this year, but the parents are having a hard time protecting the young from the eagles and other predators." An exhibit at the Cranberry Discovery Center museum in nearby Warrens states that a one-acre cranberry marsh requires about seven acres of supporting land and water. The result is a protected habitat for wildlife.

Ed Sr. stops the truck at the edge of several small squares planted with various Grygleski hybrids, flags at the edges marked "GHI x 35." "We first plant new hybrids in four-foot-by-four-foot squares," he explains, as we climb down a sandbank to the level of the vines. "After three or four years, if we feel they're showing promise, we try growing them on a ten-by-twenty-foot rectangle. If they look marketable, we plant a half acre. We don't offer mowings until they've gone through these three steps."

He bends down to loosely lift greening leaves up for inspection. This particular vine is a cross between the Howes (the heirloom variety originally from Cape Cod) and the Grygleski GI hybrid. Then he moves on to another bed and lifts up part of a vine. This is another Grygleski hybrid that he and his son are crossing with the GI. It's the berry they are most excited about. I ask him what in particular appeals to him about this vine at this time. "It has a blush," he answers from years of experience. "Sometimes you get an impression."

It takes a total of fifteen years before a hybrid is ready to be mowed and marketed. Color is important, resistance to insects and weather are important, but hybrids are bred primarily for their higher yield. In 2012 close to fifteen million barrels of cranberries were harvested from U.S. bogs. That's still not enough to satisfy the world demand for dried berries and juice. Stevens is the berry most commonly grown in Wisconsin. The average yield for the Stevens is about 360 barrels per marsh. Ed is hoping a hybrid he refers to as BG, for Beckwith-Grygleski, will yield more than 500 barrels per acre.

Ed doesn't consider himself a scientist. Since he first began experimenting with new hybrids, he has relied on the expertise of research horticulturists at the University of Wisconsin, specifically Dr. Malcolm Dana and Dr. Donald Boone. The Grygleski family, the university, and the community of cranberry growers all benefit from the results.

The earliest cultivators divided the *Vaccinium macrocarpon* Alt into three different types, based on their shape—the cherry, the bell, and the bugle. The cherry is round and darker in color than the other two. The heirloom variety named Early Black that was first cultivated by Cyrus Cahoon on Cape Cod is considered a cherry. The bell is narrower at the top than the cherry, not as dark, and with a waxy skin less liable to bruise than other varieties. Howes are bells. The bugle is more elongated, egg-shaped, paler, and less frequently cultivated than the other two. All three have been used in research for hybridization.

We walk over to a nearby set of beds and Ed explains that these hybrid plots were developed by Dr. Brent McCown and Eric Zeldin at the University of Wisconsin's Department of Horticulture. In addition to higher-yielding hybrids, the Grygleskis are hoping to develop plants with healthy vines, good fruit-budding qualities, a deep-red berry color, and the ability to hold up well. Many of these attributes derive from breeding heirloom varieties with newer varieties.

When I ask about the bed marked GH1 x 35, Ed identifies it as a number 35 hybrid pollen crossed with a GH1 hybrid variety. "It looks like a late ripening berry," he adds, "so it may be a good fresh fruit variety."

On the hundred acres of marsh they farm, the Grygleskis employ five full-time workers. In 2010 they completed installation of a new computer-operated irrigation system with doors that open automatically to let out excess heat buildup from the diesel engine. For the time being, they have no plans to convert to pop-ups. But on the new bogs they are building on land across the street from the breeding area, they are installing drainage tiles similar to those installed by Keith Mann in Massachusetts. I ask if they have any problems in Wisconsin with permitting, and Ed states unequivocally, "It's a nightmare!" An East Coast grower would remind him that he and other Wisconsin growers have more undeveloped land to work with than is available on the East Coast.

As we drive through the rest of the marsh, we spot one of the resident bald eagles and a pair of swans with cygnets. Ed points to a shrub I have noticed growing throughout the marsh. He stops so that I can get a better look. "See that," he says. "A glossy buckthorn. That's my biggest headache right now." He explains that this particular invasive plant is not only gobbling up space but that it has a growing season paralleling that of the cranberry. When the migrant bees arrive, both buckthorn and cranberries are ripe for pollination. The bees fly right over to pollinate the flowers of the glossy buckthorn, a sweet and lovely treat for a bee. Ed speculates that the drooping crane heads of the cranberry look like too much work, for a honeybee, and thus are not as appealing to a bee with choices. Even in Eden, it was difficult to control the wildlife.

Before 2012, when I asked cranberry growers if they had seen changes in the growing season that could be attributed to global warming, I received a resounding "No." I ask Ed.

"When I was little," he replies, "we would already have ice forming on our beds at night in early October. The temperatures could fall to five or ten degrees above zero. You would wait all day, and the ice still wouldn't have thawed off. The fear was always: am I going to be able to get my crop in? We can now harvest later than we ever could before." And that places even more berries in the marketplace.

Abbott Lee in the Pine Barrens area of New Jersey is less concerned about the possibility of an oversupply and more concerned about the purity of the cranberry hybrids he provides to growers. His great-grandfather was the first to grow cranberries commercially in New Jersey. His brother, Stephen, grows and harvests cranberries on the family's 2,100 acres of bogs. Abbott is the pioneer in producing foundation-level cranberries for the industry.

The first week in March and one week after twenty-five inches of snow had fallen, I visited Abbott at his plant, aptly named Integrity

Propagation. This is where he grows the hybrids developed by Dr. Nicolai (Nick) Vorsa, a plant geneticist at the New Jersey Agricultural Experiment Station. And this is the source for Keith Mann's Mullica Queens.

The temperature is thirty-seven degrees, spitting rain and snow. Despite the weather, pots of dormant year-old vines have been moved out of the protective climate of a greenhouse and are acclimatizing outside under tarps. "When purchased by a grower and moved to a new location, these plants are hardier and less apt to go into shock than a plant straight from the greenhouse," Abbott explains. He pulls open a greenhouse door, pushing it just far enough back so that we can enter without getting caught on the heating and ventilating unit protruding from its exterior. Inside the temperature is maintained at a maximum of eighty-five degrees, with a humidity level around 90 percent. Reddish brown vines cover the soil in a sea of white plastic seven-inch pots on hip-height tables. The plants will be here until the middle of June. Then they're going to start to grow.

If the plant puts out fruit buds or flowers, they will literally be "nipped in the bud." If left to bloom, flowers would yield buds; the buds yield berries; the berries yield seeds; and the seeds yield new individuals, or what Abbott perceives as a threat to future purity of the breed. After the flowers and buds are picked off and the plants fertilized and weeded, the pots are hung outside on a structure of elevated metal bars about eight feet above the ground. Electronic probes or emitters attached to the metal bars monitor the humidity and soil conditions in each pot. If the soil in a pot is too dry or too damp for ideal growth, the emitter will send a signal to a computer in the office.

By the end of September the vines will have grown in straight runners almost to the ground. At the end of October or beginning of November, the hanging baskets are taken down and moved back inside to the climate-controlled greenhouses. A month later they are harvested. Workers cut the vines, secure the bundles of vines with rubber bands, remove any dry or brown ones, and transport them on golf carts to the central assembly building. There other workers inspect them under light, remove brown or damaged leaves, straighten the vines, put them in plastic bags, add water to keep the humidity at 100 percent, and place the bags in cold storage until they are purchased or cut for new growth. Once a year the New Jersey Department of Agriculture tests each plant for viruses that could spread to other areas of the country and other marshes, bogs, and beds where the plants are to be delivered.

Abbott unlocks the door to the cold-storage room, then pulls a three-foot clear plastic bag out of one of the bins. Two stripes of green arrows, the markings used to identify Crimson Queens, run the length of the bag. A color-coded panel on the front of the bin indicates the greenhouse where the vines were grown, the date the vines were cut, and their age. From the cold-storage room, Abbott and I pass beneath a green metal mixer. Overhead, Bill Campbell, facilities manager, is emptying peat moss, vermiculite, and what Abbott refers to as "some other secret ingredients" into the mixer. Once blended, the new potting medium flows from the mixer down into color-coded foot-square flats. A conveyor belt sends the flats to a stamping machine that punches a series of holes in the soil, then transports the flats to ten men and women seated on either side of the moving belt. Rhythmically, they pick up cut vines and place them in the prepunched holes as the flats move by. From the conveyor belt the newly planted cuttings are moved outside to greenhouses to become new stock.

Sarah Bell, on the other side of the room, is taking Crimson Queen vines out of a cold-storage bag, cutting them into two-inch-long bunches, wrapping the bunches in a green plastic holder and packing them in plastic bins for an April delivery. Eight days of heavy snow made roads impassable for the assembly crew and caused Integrity Propagation to experience its first backlog of orders. The job of handling the backlog falls to Marianne Anderson, Bill's sister and Integrity Propagation's facilities director. Sitting in a windowed room off the assembly area, she can be seen poring over a ledger. "Marianne handles all the books. Orders, shipments, quality control . . . she does it all. With the two of them here, I don't even have to be present," Abbott insists.

Both Ed Grygleski and Dr. Nicolai Vorsa are creating new hybrids for the industry, but their methods and philosophy are markedly different. Ed takes one breed and crosses it with another breed to produce a new breed. "Nick" Vorsa identifies and records the DNA of individual cranberry strains. Then he isolates the strongest properties within one breed and crosses them with what he has isolated as the strongest properties of another breed.

When Dr. Vorsa began working at the Rutgers Blueberry and Cranberry Research Center, he was given strains of cranberry varieties by the New Jersey Department of Agriculture. "Nick began, and is still, sorting out the plants in an effort to isolate superior parents and produce superior progeny," Abbott says. Over a twenty-year period, Nick and Richard Novy, his graduate student, developed a DNA fingerprinting technique for cranberries. "It was very time consuming," Abbott continues, "but Nick is very patient and very conservative." He conducted eight years of field trials before releasing the new varieties. "It wasn't until some growers put pressure on Nick, saying, 'look, you've gotta release them,' but Nick knew he was working with a crop where his legacy was going to depend on the few releases he made. It's not like a rice breeder or something, where you come up with a new generation every two years."

Traditionally, growers took prunings or mowings from one bog, then cut the vines and planted them on another bog, the way Barry Paquin was planting his bog in Carver, Massachusetts. "The problem," Abbott explains, "is that you have seeds. And each of those seeds is a unique individual." The seed, when planted on a cranberry bog, germinates. If it survives, it is probably "vegetatively superior." That means that it wants to produce runners, not fruit. A cranberry vine that produces fruit probably doesn't produce vegetation and vice versa. Mowings taken from an area of one bog could consist of many different individuals, some vegetatively superior and some less so. From a geneticist's point of view, this leads to genetic deterioration.

Ocean Spray contributed over $150,000 to Nick's research. The growers, who were paying for the work, felt they deserved to see results. Finally, Nick released Crimson Queen and DeMoranville, named after Irving DeMoranville, the former director of the Massachusetts Experiment Station, whom Nick credits with initiating hybridization to the industry. "I crossed Ben Lear with Franklin, a variety released by Irving DeMoranville in the 1930s," Nick explains. Two years after the first releases, Nick released Mullica Queen, the variety planted by Keith Mann.

The crosses for DeMoranville and Crimson Queen were made in 1998. The plants were released in 2004 and 2005. The cross for Mullica Queen was made in 1997. "No. 35" was another of Irving DeMoranville's hybrids, but it was never released due to its relatively white color. Phil Marucci, namesake for the Phillip E. Marucci Center for Blueberry and Cranberry Research, where Nick is the director, kept a few No. 35 berries and grew them on a small plot on the bog across the street from the station. In trials with No. 35, the plant yielded the highest number of berries in the ten-year program. The berries were smaller than Stevens, but they set far more fruit.

Crimson Queen showed a yield of 628 barrels per acre when grown at the Rutgers Experiment Station; Mullica Queen, 637 barrels per acre; and DeMoranville, more than 400. Those higher yields, together with hardiness and resistance to pests, are what the growers want. But how do you go from Nick's tiny plot of Crimson Queen to ten acres or a hundred acres?

Abbott proposed Integrity Propagation as the next step. In 2002 he and Nick went to the various deans of Rutgers University and floated their concept for transitioning to production of the new varieties. Rutgers liked it. Abbott insisted that if he was going to invest his money in the new venture, he would need to have exclusive rights to the distribution to recoup his investment. Rutgers countered with three requirements: that Abbott keep the price reasonable, keep the varieties pure, and meet the demand. A year and a half later Abbott and Nick had a final written agreement, and as of 2014 they have met all three requirements.

Across the street from Nick's office, twelve acres are available to him for the cranberry-breeding program. He cycles about three thousand new seedlings from crosses every year. He also takes out three thousand new seedlings each year. Each plot begins with about twenty-four plants. It takes two to three years for that plot to get established and produce mature berries, then another four years to evaluate the yields, determine the anthocyanin levels, and judge the variety's resistance to rot. After six or seven years Nick selects those plants with the highest levels in all three categories. He then tears up that field to make room for a new generation. "We've just released a fourth variety—the NJ9520-20 until a better name is found. It's a really nice fruit. I can honestly say that it has the highest color of any cranberry variety. And," Nick adds, "it should have the highest anthocyanin levels available."

The anthocyanin level (a measure of antioxidants) for DeMoranville is approximately four times higher than the level for Stevens, the most frequently cultivated cranberry. Abbott believes that the De-

Moranville color is also stronger. The hybrid produces a larger berry and it flowers earlier, enabling the berries to have a longer growing season even in the Northeast. Supply and demand still rule. Ocean Spray is still projecting a price of sixty to sixty-five dollars per barrel, but the danger of oversupply is always present, and as these higher-yielding berries are being harvested, the supply will increase. One of the growers who bought and harvested Crimson Queen reported a yield of 620 barrels per acre. His Stevens came in at 385 barrels per acre. And that's a high yield for Stevens, formerly the high-yield variety, whose larger size translates to a higher yield as measured by volume. "These new varieties pay for themselves very quickly," Nick points out. "I was very pleased when the numbers started to come in . . . that we definitely did improve them genetically."

No. 35 is a cross between the Massachusetts heirloom Howes and the Wisconsin heirloom Searles. Nick crossed No. 35 with LeMunion, named after Norman LeMunion, a New Jersey grower who discovered the berry growing in the New Jersey Mullica River Watershed. "It's the first variety with bloodlines in Massachusetts, New Jersey, and Wisconsin," Nick states. In making the cross, he was looking for high yield and anthocyanin color.

The higher yields complicate the already existing problem of oversupply. To keep a sustainable balance and not allow a large yield to depress prices, the Department of Agriculture's Cranberry Marketing Committee has to be creative. And they are. Cranberry doughnuts are now available in Japan; organic fresh cranberries at Sainsbury, one of England's larger food chains; chicken sandwiches with cranberry sauce are sold in Korea; and cranberry smoothies are made in Germany. The percentage of foreign sales rose to 25 percent of the overall market for U.S. cranberries in 2010 and is projected to reach 33 percent in 2015.

"When these high-yielding varieties were released," Nick says, "all the test plots showed that they were good varieties, doing well. But they hadn't been tested on larger acreage. So the first growers to plant these varieties took a risk. Now that they [the plants] are performing the way we had hoped they would, now the growers coming into it may have to pay a higher price because the real risk was born by the pioneers."

Nick would like to eventually map the entire cranberry genome. Given the gargantuan number of genetic possibilities, it's not possible for him to plant a trillion or so genetic crosses on the twelve acres across the street. The greater the genetic data he gathers, the easier it will be to isolate the desired quantities, making for a more efficient breeding program.

Initially, he traveled to Oregon, Washington, British Columbia, Wisconsin, Massachusetts, New Jersey, and as many places as he could to find producing cranberry bogs, marshes, or farms. From each, bed, bog, or marsh, he collected samples and brought them all back to his lab. Some newer varieties such as Stevens turned out to be pretty consistent in their DNA. Others such as Searles, even though they were considered as one variety, were made up of five or more varieties. One reason could be that when cranberry bogs were renovated or replanted with a higher-yield variety, the old variety was never wholly destroyed. Fulfilling its role to increase the species, and missing a more appropriate partner, it propagated with the new variety.

Abbott Lee and Nick have an agreement with Acultra Propagator, a Wisconsin firm. Acultra buys stollons from Abbott, plants a field of them, then mows them and sells the mown vines. Acultra's customers can then plant the mowings, using Barry Paquin's method, and Acultra keeps the original plants growing for new mowings. Acultra is the only propagator authorized to sell Nick and Abbott's vines.

"We hope to keep these varieties as clean as possible for as long as possible," Nick says. He goes out to Wisconsin three times a year to inspect Acultra's beds and expects that for the next six to ten years, the beds will remain uncontaminated by outside varieties. The Acultra vines are not considered foundation level, to differentiate them from the pure vines raised under controlled conditions at Integrity Propagation.

John Decas and other independent growers, however, voice some issues with the distribution of Nick Vorsa's hybrids. The idea that a portion of the public was prohibited from acquiring the high-yield results of publicly funded research struck a discordant note in an industry traditionally noted for sharing. "Then we find out that millions of dollars of public money has been spent to develop these superfruit. And we're not allowed to benefit from the results because we're not part of the Ocean Spray cooperative?" John's voice rises in disbelief.

In 2011 Rutgers offered to sell the new hybrids to independent growers, but only if they pay an upfront fee over and above what an Ocean Spray grower has to pay. The independents claim that the Ocean Spray growers have had a five-year run using taxpayer's money while independents were not "invited to the party" and that advantage should more than make up for Ocean Spray's providing initial seed money.

"You know," John points out, "the original breed stock originated right here in Massachusetts, from the University of Massachusetts Experiment Station. We've never charged for that." When he was a student at the station, John did his internship under Irving DeMoranville, the namesake for Integrity Propagation's DeMoranville hybrid, and he is offended by being denied the right to grow a variety named after his professor, while people who don't even know who

Irving DeMoranville was are allowed to purchase the vines bearing his name.

Nick Vorsa prefers to remove himself from the argument and focus on his research. In addition to his work with plant genetics, he also studies the health benefits of cranberries. Along with Catherine Neto in Massachusetts, he is looking into the relationship between proanthocyanidins (PACS) and cancer treatment, but with a slightly different focus. Dr. Vorsa and Ajay P. Singh, another Rutgers scientist, have found that when they have extracted PACS in cranberries, the resultant compound causes a 600 percent increase in the effectiveness of platinum-based drugs in the treatment of ovarian cancers. Nick links the PACS in cranberries with how they are dispersed. As he speaks, his reverence and awe of nature are contagious. He highlights that each living species exists to continue the species. If you bite into a green banana or an unripe persimmon, the taste is unpalatable. Other animals have the same reaction. That's partly due to the PACS in the fruit warning us, "I'm not ready to be eaten yet." When they are ready, the PAC chemical compounds diminish. In cranberries the PAC level remains strong. Most fruit other than cranberries rely on animals for dispersal to new growing areas. But the system works only if the fruit is ripe and ready to seed. Deer will eat cranberries, but they crush the seeds in their teeth instead of eliminating them whole. Fortunately for us, the cranberry is waterborne, rather than animal borne. Thus, it doesn't have to taste sweet to attract animals. It only needs to float. Thanks to its internal air sacs, it can and does.

Cranberry acidity levels are 2.1 to 2.2. Blueberries have an acidity level of 0.5, and blueberry sugar levels are higher. The red color of cranberries, blue color of blueberries, and reddish purple of grapes are indications of anthocyanins. But, the taste of blueberries, grapes, cherries, and most other palatable fruits comes from a particular chemical associated with a pleasant taste. The chemical in cranberries is an antiseptic—another reason to believe the cranberry was not designed to rely on animal dispersal, nor was it designed to be eaten by humans. For it to be palatable, it had to be converted from tart to sweet.

*We require just so much acid as the cranberries afford in the spring.
No tarts that I ever tasted at any table possessed such a refreshing,
cheering, encouraging acid that literally put the heart in you and
set you on edge for this world's experiences, bracing the spirit, as the
cranberries I have picked in the meadows in the spring. They cut
the winter's phlegm, and now you can swallow another year of the
world without other sauce.*

—Henry David Thoreau, *Wild Fruits*

12

FROM TART TO SWEET

If you've seen cranberry growers, they can pick up a handful of cranberries off the vine and eat them. And that's about the way cranberry juice cocktail tasted, so we did the not so brilliant thing of putting more water and more sugar in it. It then became a very palatable drink, and we initially marketed it across the country. And that really started to turn the business around.

—Edward Gelsthorpe, past president of
Ocean Spray, AdEast interview, 1974

They're in your morning cereal, on your salad, and in your trail mix. They're organic at the health food store or traditionally harvested at your grocery store. They're cranberries, but they're not tart. In the raw state, a cup of cranberries contains only 50 calories, no fat, and virtually no cholesterol. When not enhanced, a cranberry has little sodium and lots of fiber, plus Vitamin C—a nutritionist's dream fruit. So why transform the cranberry from tart to sweet?

To support larger brains and the greater number of children their bodies produced, our hunter-gatherer ancestors needed more fat than most other primates. Because food wasn't readily available, early *Homo sapiens* needed to store extra fat to survive periods of scarcity. To provide that energy, they had to burn the small amounts of sugar they found in the available fruits and sweet vegetables. Ever adaptive, our bodies learned to convert excess sugars to fat. Voila! Energy requirements and the body's need for fat were both met by satisfying the resultant craving for sugar.

We humans still crave sugar, but in most countries today sugar is readily available in many forms, and that availability eliminates the need to burn calories to find it. Unless we undergo another adaptation, optimum human health will continue to depend on the perfect balance of sugar intake to calories. An imbalance results in obesity and diabetes, problems we face as manufacturers add more and more sugar to foods to appeal to our sugar craving.

A twelve-ounce bottle of Ocean Spray Cranberry Juice Cocktail contains forty-five grams of sugar; the same size glass of "100% Juice" by Ocean Spray, fifty-two grams; "Light Cranberry Juice," sixteen grams; and diet, less than one gram, the largest ingredient being water. The same twelve-ounce bottle of Coca Cola, long seen by nutritionists as one of the prime sugar drinks responsible for diabetes and obesity, contains thirty-nine grams of sugar. So how do you decide whether or not to purchase cranberry products for you and your children?

In the United States, urinary tract infections are the second most common reason for infectious disease hospitalization in people over sixty-five, and Medicare costs alone amount to more than $1.4 billion each year. The economic and societal costs of obesity are significantly higher. So what's the solution? For one, it is simple to make your own cranberry juice at home. That way you can gauge how little or how much sugar to add, and slowly lower the sugar content as your taste buds adapt (see recipe in the section "For the Cook").

If you don't choose to spend the time preparing your own juice, there are some bottled, undiluted cranberry juices found at health food stores or in the natural food departments of most large grocery stores. Knudsen is one brand. Trader Joe's has another. These juices take a little getting used to, and for many of us the homemade variety tastes a lot better.

Ironically, a warming planet may provide a solution. Cranberries on the West Coast are inherently sweeter mainly because they have a longer growing season. If the trend to warmer winters continues in Wisconsin, New Jersey, and Massachusetts, areas where the bulk of the cranberries are cultivated for juice, the growing season could also be longer, up to a tipping point. The berries will then be baked by the sun for a longer period of time, producing naturally sweeter berries. Until that happens, make your own cranberry juice with less sugar and reestablish the balance for healthy eating.

For those who prefer to get their cranberry health benefits from sweetened dried berries, Decas now offers two new Paradise Meadow lines: LeanCrans for adults, and for children the Funny Face dried cranberries line. The two new offerings contain half the sugar and 25 percent fewer calories than regular sweetened dried cranberries. They also boast that they provide five times the fiber.

John Decas and I are sitting in a small conference room within a large building the length and width of a small city block. In 1996 it was an empty field. "We initially took a gamble on our sweetened dried cranberries," John explains. "Ocean Spray had a patent on their technology, and the challenge for Decas was to make a better product than the Ocean Spray product while not encroaching on the cooperative's patent. So we hired food scientists, patent lawyers, and put together a team of highly dedicated, motivated, intelligent people, trained in all the things that go into producing a quality product. Together, we came up with what we felt was a better product than Ocean Spray's. And," he adds, "indeed it is."

The Decas team wanted a product less sweet, with more cranberry than sugar. On the packaging for their new product they wanted to find a way to list cranberries before sugar in the list of ingredients.

"Ocean Spray, at the time, extracted all the juice they could out of their product because they were essentially a juice company," John maintains. "When they removed most of the juice, they also removed natural acids, so they had to replace that with artificial acid to give it some kind of a cranberry tang."

When Decas came out with its dried cranberry product, Ocean Spray challenged them on the labeling. The Department of Agriculture ruled in Decas's favor, allowing the company to list cranberries ahead of sugar on its labels. Ocean Spray subsequently retooled and began producing a dried cranberry product with less sugar. "So in a sense we were kind of the pioneers for the modern sweetened dried cranberry," John says in retrospect.

Ocean Spray takes a different view. "In the mid '90s," CEO Randy Pappalardo stated, "we'd extract juice from the cranberry and then leave behind what we call the hulls. And we paid people to come and haul away the hulls. Our first breakthrough was realizing that by reinfusing a little bit of cranberry juice back into that cranberry hull, we could create something that we now call a sweet and dried cranberry, and we've branded it as Craisins. . . . And of course, everyone loves Craisins."

Ocean Spray Cranberries Inc. is a branded company, a cooperative of growers, that packages retail products under the Ocean Spray label. A division of the cooperative sells cranberry concentrate, sweetened dried cranberries, and other cranberry derivatives used in their corporate customer's products. To the public Ocean Spray and cranberries are synonymous. In fact, only about 70 percent of industry sales are generated by Ocean Spray growers. Of that number, 70 percent or more is sold as dried fruit, the remainder as concentrate. John Decas thinks the cooperative may have created a number of conflicts for itself. "They sell concentrate to their competitors. So what's anyone going to do with concentrate? They're going to make juice out of it, and with that juice they compete with the Ocean Spray brand."

The concept of removing water from the berry was not new. During World War II John Makepeace, whose family company was the world's largest cranberry producer, dried his whole crop in 1944 and sent it overseas to the military. The label read, "Makepeace CRANNIES, Dehydrated Cape Cod Cranberries." Ocean Spray also shipped blocks of dehydrated or evaporated cranberries to the armed forces. One brick, four by four by three inches, would make about a hundred servings of cranberry sauce after the addition of water and sugar. Enough of these blocks were shipped to warrant Ocean Spray's receiving the A award for agriculture from the U.S. government.

John Decas is quick to admit that the rest of the industry benefits from Ocean Spray's larger size and pooled income. "They advertise. They promote. It helps all of us. Their research on new products also helps all of us. I like to think our research and new product development also benefits the marketplace and therefore the industry."

He maintains that his sweetened dried berries retain their natural acid content. And the Decas company is happy to enhance the basic berry for a specific market. They coat the berries with cinnamon for the UK market, slice them for the bagel market, dice them for the cookie market, flavor them with orange or blueberry for the cereal market, infuse them with moisture for the bakery market, and chop them for power bars.

Chances are good that any retail package of sweetened dried cranberries without an Ocean Spray label is from the Decas company. Under its Paradise Meadow label, the company has just introduced a dried berry with 50 percent less sugar and one-third fewer calories for the customer concerned with weight and diabetes. It also offers an organic sweetened dried berry. When Decas built the plant we are in, it was engineered to produce five million pounds of sweetened dried cranberries per day. Today the company handles about five hundred thousand barrels of cranberries or fifty million pounds each year from Massachusetts, Wisconsin, and Canada. Its manufacturing plant is running twenty-four hours a day, seven days a week, and the five million pounds per day plant capacity is now up to twenty-five million pounds. "The work never stops," John says. He recently put in a new berry dryer and is five to six weeks behind in filling orders. "And so is Ocean Spray," he adds.

I describe the plastic boxes of sweetened dried cranberries with the label of the store where I shop, and I ask him if they are his berries.

"Are they good?" he asks.

When I respond positively, he laughs, "Then they're ours."

John Decas is presently running clinical studies in China, India, and the Czech Republic on the health benefits of eating cranberries. To counteract the image of farmers in a niche business and to ensure credibility of the test results, he is careful to use well-respected scientists and the peer review process. "For example," he explains, "elderly women are prone to urinary tract infections. Studies show that if they drink a glass of cranberry juice every day, they're less likely to come down with the infection. So in effect what you're doing is reducing people's dependence on antibiotics. And, the overuse of antibiotics has become a big problem in the world." Accordingly, the results of a 2009 study by scientists at the University of Dundee in Scotland showed that a low dosage of antibiotics in preventing recurrence of urinary tract infections had only a limited advantage when compared with the effects of a cranberry extract.

Native lore on the cranberry's ability to combat urinary tract infections generally focused on women, traditionally the group that contract urinary tract infections more frequently than men or children, but a 2009 study by a group of Italian scientists found that a regular diet of concentrated cranberry juice also tended to prevent a recurrence of the symptoms of urinary tract infections in children. Other studies on the ability of cranberry proanthocyanidins (PACs) to reduce the incidence and severity of urinary tract infections include a British study conducted on men with a high risk of prostate cancer. Results of that study indicate that cranberries may lessen the incidence of urinary difficulties in older men. Ongoing Decas studies attempt to see if the PACs in cranberries can be extracted and injected into other fruit while providing the same health benefits.

Additional research on the cranberry's role is ongoing in the areas of cardiovascular disease, various cancers, and cognitive function. Cranberries have higher levels of antioxidant polyphenols (373 per half cup) than other fruits such as blueberries (181), red grapes (296), or cherries (231). Cardiovascular disease is responsible for more deaths in the United States than the next five top diseases. Researchers at the Jean Mayer USDA Nutrition Research Center on Aging at Tufts University found that the polyphenols and flavonoids in cranberries may help reduce the risk of cardiovascular disease by increasing the resistance of LDL to oxidation, blocking platelets from clumping together, and lowering blood pressure. More than a million deaths are related to cardiovascular disease, and the cost of treating it is over \$350 billion per year.

Breast cancer, the form of cancer that attacks more women than any other form, causes the highest number of deaths due to cancer in women. Researchers at Cornell University studied the reputed anticarcinogenic qualities of cranberries in an attempt to understand if the reports were valid, and if so why. Their findings showed that a small dose of the phytochemical extracts in cranberries significantly blocked further growth of breast cancer cells for four hours, and after continued dosages were given for twenty-four hours, growth of the cancer cells was approximately six times lower than that of untreated cells. They concluded that properties in cranberries are capable of suppressing growth of breast cancer cells.

Another area presently being studied relative to the value of cranberries on human health is the field of antiaging. Alzheimer's disease and its more forgiving cousin, senior dementia, place a huge burden not only on those suffering from the disease and its counterpart but also on their family members. In both cases those afflicted suffer from various degrees of memory loss and loss of coordination. At least two scientists, Barbara Shukitt-Hale and James Joseph at Tufts University, are engaged in research based on preliminary findings showing that the high antioxidant levels in cranberries may help protect brain cells from damage leading to the loss of motor and cognitive function. At UMass Dartmouth, Catherine Neto's group and the research lab of Maolin Guo have received funding from the National Institute of Health and the American Parkinson Disease Association for similar research on the cranberry and stress reduction leading to improvement in cognitive impairment.

As Dr. Nicolai Vorsa reminds us, the cranberry's bitter taste makes it repugnant to most animals. Thus, it resists being animal borne, enabling growers to control its purity, habitat, and safety to humans. A more perfect fruit would be hard to design. It just so happens that conditions in the northern latitudes of North America have been favorable to its growth. Let's hope they remain so.

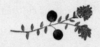

*If it looks like the last day of the world is upon us and the end of
life may be coming . . . and you realize this moment while planting
trees, well, don't stop planting.*

> —Arabic proverb by Aziz Bousfiha, in Gary
> Paul Nabhan, *Growing Food in a Hotter,
> Drier Land*, 2013

13

WARMING OR CHILLING

Do not think, then, that the fruits of New England are mean and insignificant while those of some foreign land are noble and memorable. Our own, whatever they may be, are far more important to us than any others can be. They educate us and fit us to live here.

—Henry David Thoreau, *Wild Fruits*, 1851

Flying toward the Big Island of Hawaii from the East Coast, our plane had been shrouded in clouds for what seemed like an interminable amount of time. I was tired and hungry, and my bones, muscles, and joints had been forced to conform to a twenty-eight-by-twenty-eight-inch-wide space for approximately ten hours. Then, as I looked out into the dark, I saw a pinnacle of rock sticking up above the clouds, promising an end and a beginning to the trip. This was my introduction to Mauna Loa.

Since the 1950s, atmospheric data has been continuously monitored and collected at the Mauna Loa Observatory, 11,135 feet above sea level. This data indicates a trend, the Keeling Curve, showing that carbon dioxide in the air has increased by approximately 0.9 percent per year, spiking upward during the past fifty years. During that same period the data from stations on land, sea, and satellites show a parallel warming trend. Meanwhile, the mean annual temperature in the northeastern states has increased by more than 1.5 degrees, and each of the past three decades has been warmer than the preceding one. Despite scientific data, a debate still rages on whether or not the warming trend is due to humans or nature, but there is no denying that plants and animals, unmindful of the debate, are changing and adapting to a world different from the one they have known for the past hundred years. Birds are migrating north earlier in the spring and, in the case of coastal seabirds, flying impossibly long distances because the insects they rely on for food are moving to higher latitudes and elevations. A 2009 report issued by the Audubon Society indicates that over the past forty years, almost 60 percent of 305 North American bird species have gradually shifted their ranges to the north by an average of 35 miles. The purple finch, a bird regularly seen at backyard bird feeders, used to migrate farther south in the winter. As temperatures have risen, it has shifted its overall range 433 miles to the north.

Shifting zones for plant hardiness are also a measurable sign of climate change. Since 2011 five hundred million trees have died in the Southwest due to drought, fire, heat, and disease, probably due to aspects of a warmer environment. Acknowledging that home gardeners need a guide to show them what plants can safely be grown in their climate zone, the U.S. Department of Agriculture commissioned the University of Oregon to revise the *Plant Hardiness Zone Map*, last updated in 1990. The new map was completed in 2009 and finally released in January 2012.

Meanwhile, plant biologists are observing flowering times for wild plants and comparing them to herbarium specimens collected a hundred years ago. Their observations are indicating that phenological events, such as the timing of nature's cycles of flowering, are occurring earlier due to increased temperatures associated with global warming.

While significant work has been done on the relationship between warmer temperatures and wild species, far less research has been devoted to the relationship of cultivated species to warming temperatures. *Vaccinium macrocarpon* Alt, our cultivated cranberry, is particularly vulnerable to the effects of climate change, especially if earlier flowering destroys the fragile balance the fruit enjoys with many of its insect predators and pollinators.

In 1863 President Lincoln created the Department of Agriculture. Land grant colleges, largely agricultural in focus, followed. Until then, farmers had no way of keeping up with the newest developments in insect control, fertilizers, and irrigation other than by word of mouth and the occasional trade newsletter. To fill this void, individual stations within the larger colleges of the University of Michigan, Rutgers, the University of Massachusetts, and others were created with a focus on the particular state crop that employed the largest number of people in each state: blueberries and cranberries in New Jersey and cranberries in Massachusetts.

The UMass Cranberry Experiment Station has been at the same Wareham location in the heart of Plymouth County since 1910. Its focus is research and education. The building housing the office of Carolyn DeMoranville, the station manager, has a distinctly nineteenth-century feel to it. Walls are pine paneled, darkened over the years. Furnishings include the oak rolltop desk Carolyn's father sat at when he had the job she has today.

She greets me wearing a no-nonsense outfit—loose flannel shirt over T-shirt tucked into trousers. "Our research is related to growing cranberries, not uses for cranberries," she tells me, in a matter-of-fact manner. In other words, the station does not do research on the health benefits of cranberries. That's left to Catherine Neto and her University of Massachusetts Dartmouth graduate students.

A student working toward completion of a PhD in insects, plant diseases, plant nutrition, or weed management might come to the

Wareham station to work with its faculty. A cranberry grower could stop by to pick up a copy of the *Cranberry Chart Book* for Massachusetts to learn about the region's newest insect visitors and how best to control them. Or the grower might take home a pamphlet titled *Planting Cranberry Beds: Recommendations and Management, Cranberry Water Use, an Information Fact Sheet, Cranberry Irrigation Management, Best Management Practices Guide for Massachusetts Cranberry Production, Frost Protection for Massachusetts Cranberry Production,* or *Nitrogen for Bearing Cranberries in North America.*

Carolyn's office is located in a chamfered-stone building. Downhill from the building, adjacent cranberry bogs enable scientists and students in residence to conduct on-site research. The day I visited, Nick Vorsa's hybrid, DeMoranville, back at the home of its namesake, was being grown on several of the station's experimental bogs.

Behind the station four brown bogs have previously been de-vined in preparation for an upcoming experiment. These bogs go back to the 1800s, when the land and buildings were privately owned. Carolyn intends to plant about half of one bog with the Stevens variety, a popular hybrid produced for its high yield. The remainder of the bog will be planted with approximately twenty-five different varieties, several sections with the same one. This duplication will allow different tests to be conducted to determine how a particular berry fares under different conditions. Plants will range from heirlooms to varieties yet to be named. The latter will allow growers to see what the new plants look like, how they thrive in Massachusetts conditions, and how their productivity rates compare to those of the older varieties. And it will spare a grower having to devote a producing bog or a new bog to an untested experiment.

On four additional strips, each forty feet by eight feet, Carolyn will plant the current Massachusetts industry standards: Early Blacks and Howes. Next to them, she plans a similar arrangement for the newest varieties, many grown by Ed Grygleski in Wisconsin and Nick Vorsa in New Jersey. These sections are to be subdivided into smaller areas to allow researchers to conduct comparative experiments to determine if the station needs to modify its pest-management recommendations.

Quebec and Wisconsin have different climate conditions and different seasonal insects from those in Massachusetts. New Jersey, whose climate and proximity to the coast more closely resemble that of Massachusetts, is home to the cranberry fruit worm, considered by Massachusetts growers to be the pest capable of causing the most damage. In New Jersey the fruit worm's emergence coincides with that of the blueberry, not the cranberry, so research isn't abundant on how the New Jersey varieties might fare in the Massachusetts horticultural zone. "And that's the kind of thing we want to be able to do," Carolyn says. "We want to make the 'farm' so it's very productive, modern, cutting edge, with all the most up-to-date things installed so that growers can come here and see how it might work for them."

Up a slight grade from Carolyn's office is a long, one-story building, just wide enough to accommodate a corridor with labs on each side and long enough to house a library at the far end. Here PhD students conduct research on the effectiveness of various chemicals in thwarting the latest insects and molds that thrive on the vine's roots, flowers, and berries. Funding for the research is provided by large manufacturers of chemical compounds who produce fungicides with names such as Ridomil, Dithane, and Elite 45 DF. All are competing for the job of eliminating threats to a $340 million industry responsible for approximately five thousand jobs in Massachusetts and another ten thousand in Wisconsin, New Jersey, Washington, and Oregon.

Dr. Frank Caruso, senior scientist and plant pathologist, welcomes me into his office, the first on the left as I enter the building. For more than thirty years, he and his graduate students have been recording data on the Early Black variety to determine which chemicals are the most effective in eliminating a constantly changing roster of rotters, borers, and chompers that feed on the cranberry plant. Dr. Caruso's data is of interest to me, not for the purpose he has collected it, but rather to see if it indicates a trend in flowering times and, if so, how long might it be before America's founding fruit moves north beyond our borders.

We shake hands, and he points to a gray, vertical file cabinet bursting with folders. He explains that the material I am looking for is all there, not necessarily in chronological order. I am free to take whatever I want and bring it to the library to analyze. In exchange for the privilege of using it, I am only too happy to put the material in chronological order from 1985 through 2011. I pick out an armful of folders and head for a table in the library.

The impetus for my exploration of cranberry flowering times is twofold. First is the National Arbor Day Foundation's *Hardiness Zone Map* released in 2006. The map indicates that many areas throughout the United States are one hardiness zone warmer (approximately ten degrees) than the zones shown on the U.S. Department of Agriculture's 1990 *Plant Hardiness Zone Map*. The 2006 map shows a significant northward shift in warming.

Second is a paper titled "Herbarium Specimens as a Novel Tool for Climate Change Research," published in the journal *Arnoldia*. For their research project Dr. Richard Primack and his coauthors chose 229 different living plants at Boston's Arnold Arboretum in 2002 and compared their flowering times with the dates listed for the flowering of dried specimens of the same plants over the past hundred years. During that period the average temperatures in Boston rose 2.7 degrees Fahrenheit. The Arnold Arboretum is bounded by several major highways used by commuter traffic to downtown

Boston. Thus, a small portion of the rise in temperature and a factor in flowering dates can be attributed to urbanization over the past hundred years, resulting in more paved surfaces radiating heat and fewer trees to remove some of that heat.

What Dr. Primack and his fellow scientists found, after taking into consideration years with extremely hot or extremely cold mean average temperatures, was that on average the flowering trees and other plants studied in 2002 flowered eight days earlier than the same plants had flowered at the beginning of the twentieth century.

Cranberries are, for the most part, grown in areas that are less urbanized than Boston, and it occurs to me that the *Vaccinium macrocarpon* Alt might be the perfect subject to help determine what consequences the gradual warming of the planet might have on fruit-flowering times and consequently how that might affect the relationships of the plant with the bees that pollinate them; the insects, fungi, and larger animals that threaten their survival; and ultimately their long-term viability in North America. So I begin to systematically go through Dr. Caruso's pages of data.

Once I have gathered together the various dates and percentages of flowering, I make a list of the years 1985 through 2011. Then beside each year I write down the various dates when fungicides had been applied and the corresponding notations on the percentage of flowering for each particular date, as observed by Dr. Caruso's graduate students. What I now have is a series of different percentages and different dates for each year, and I can extrapolate to create a base percentage for flowering. Fungicides, to be effective, need to be applied when close to 10 percent of a cranberry bog is in flower, so I use the dates observed to draw a graph showing the date for each year when this occurred.

To best understand what the graph is showing me, I divide it in half by years (1985 through 1998 on the left, and 1999 through 2011 on the right). Then I divide it horizontally from the lowest recorded date to the midline (May 31 through June 12) and from the midline to the highest date in the quadrant when 10 percent of the bog was flowering. The results are that from 1986 to 1999 six dates are below the line and seven above, and from 1999 to 2011 four are below the line and nine above. I decide to look at the extremes and compare the first seven years to the past seven years. As you might expect, that difference is even more pronounced. From 1985 to 1991 five are below the line and two above. In contrast, from 2005 to 2011 two are below the line and five above. This is the opposite from what I had expected.

When Dr. Primack and others compared flowering times at the Arnold Arboretum with the flowering times previously recorded for the arboretum's specimens, they found that the plants in 2003 flowered earlier than the dates indicated for flowering of the same species at the turn of the century. Before I thought it through, I expected to get the same findings from comparing flowering times for cranberries. As you can see, my results were the opposite of Dr. Primack's findings. To my untrained eye, they make it look as if the cranberry plants are flowering later, not earlier.

As previously mentioned, for the cranberry to bud, the plant needs a certain number of winter chill days (approximately thirty-two to forty-five degrees). Without the required chill days, it can't break out of its dormant period and begin to produce flowers to be pollinated and subsequently to produce berries. I surmise that maybe Frank Caruso's data, when broken down as described, is indicating that in years with warm winters, the plant will flower later because it has taken longer to accumulate days cold enough to qualify as chill days. If that is true, the real danger is that in the absence of chill days in the winter together with warmer summers, the cranberry will stop flowering and the production of cranberries will collapse in the United

States, unless growers can find a means to keep the vines cooler in the winter. A tall order.

When I describe my findings to Domingo Fernandes, the grower from Carver, Massachusetts, he tells me that for a period of at least twenty-five years, he has kept records of the dates he sprayed fungicides on some bogs he owns that are not organic. I am elated when he offers to send me the dates from his records.

By now I know I need to speak to someone who understands how to read data, and I contact Dr. Primack at Boston University. He agrees to see me, and when we meet he introduces me to the concept of regression analysis, in this case comparing the dates for 10 percent flowering with local temperatures for the same period. He kindly offers to plug my data into the temperature data he and Libby Ellwood, his graduate student at the time, have amassed. He also proposes checking the flowering dates for various herbarium specimen of *Vaccinium macrocarpon* Alt to see how they compare with today's flowering dates.

The herbarium specimens studied consist of a total of thirteen specimen, nine from the collection at the Harvard University Herbarium and four from the Boston University Herbarium. The specimens selected for the study are only those that contain some opened and some unopened flowers and thus most closely approximate an early flowering stage. All were originally collected in eastern and southeastern Massachusetts between the years 1883 and 1916.

For the Carver and Wareham data, Dr. Ellwood used mean monthly temperatures collected from a weather station at Plymouth, a town adjacent to both Wareham and Carver; for the herbarium specimens, she used temperature data recorded at the Blue Hill Observatory in Milton, one of the few weather stations with records dating over one hundred and fifty years.

The results of the study indicate that *Vaccinium macrocarpon* Alt, the cranberry that has been cultivated in southeastern Massachusetts for almost two hundred years, is influenced by spring temperatures, and for each 1.8 degree increase in May temperatures, the cranberry flowers two days earlier. Although the time span recorded in Carver and Wareham covers too short a period to strongly define a trend, when compared with the data for wild plants, recorded flowering dates for the cultivated cranberry indicate that cranberries are flowering earlier. It is safe to assume that as temperatures in the region continue to rise due to global climate change, the cranberry will flower at an increasingly earlier date.

The implications for growers are significant. John Decas now plans to protect his plants from frost in early April, when the plants break dormancy and are fragile. He recalls that he didn't used to worry about frost protection until late April into May. In addition, insects and pathogens that rely on certain plants to continue their species have also been migrating at a rate greater than two miles a year since 1960, following the migrations of their preferred preys. The result is that John and other growers, to maximize fungicide and pesticide effectiveness, will need to look for the early flowering stage at an earlier date, especially in an unusually warm year. The migrant population of honeybees that pollinate the flowers will need to arrive on the bogs earlier, and part-time employees for harvesting and processing will need to be hired sooner. If warming temperatures allow insect pests to winter over, they may increase in number or at least have a longer season to attack the cranberry. A larger number of harmful insects will require more spraying of pesticides and thus a larger expenditure on the part of the grower and additional pressure on our environment. Finally, if cranberry and insect cycles are not changing at the same rate, pollination may be negatively impacted and yield may be either reduced or eliminated.

In a recent issue of *Earthwise*, the newsletter put out by the Union of Concerned Scientists, the editors report on Pennsylvania's apple industry and predict that by 2050, if global warming emissions continue to rise at the levels they have been rising, the chill-factor temperatures required for blooming will be ensured in only 50 percent of winters. As we know, cranberries also need a finite number of chill days to produce fruit.

It is clear from the comparison with herbarium samples and the data collected from Frank Caruso and Dom Fernandes that the timing of cranberry flowering is determined by May temperatures, and with climate forecasts predicting greater warming, we can expect even earlier flowering. So what can be done to save America's founding fruit from becoming a casualty of global warming?

14

SOMETHING GAINED,
SOMETHING LOST

*We are fortunate in that the area in which cranberries can be
grown is limited. This places a natural restriction on overproduction. It also confines growers to small areas where they
can become acquainted with one another. The cranberry industry
is probably outstanding for the friendly feeling and the lack of
competition among growers.*

—President, Cranberry Canners, "Report
for the Fiscal Year Ending May 31, 1943"

In 2012, after the annual harvest, Michael Hogan, CEO of AD Makepeace, spoke from the Makepeace bogs, a thirteen-thousand-acre
holding in Wareham, Massachusetts, considered to be the world's
largest cranberry bog: "We're having warmer springs, we're having
higher incidences of pests and fungus, and we're having warmer falls
when we need to have cooler nights." It had become apparent to Mr.
Hogan and to others that cranberry profits were being eaten into not
only by cranberry fruit rot, which had become more aggressive due
to climatic changes, but also by an increase in fuel costs to keep the irrigation pumps running for more hours, the cost for extra fungicides
and extra labor to apply the fungicides, and additional technology

required to more carefully monitor temperature changes and moisture levels at the bog.

To make matters worse, the company was being paid two dollars less per barrel by Ocean Spray because of the higher percentage of fruit rot in the fruit and the whiter, less desirable color of the fruit due to fewer chilling hours. For the 370,000 barrels of cranberries Makepeace had delivered to Ocean Spray, that added up to a substantial cut in profits.

Mr. Hogan spoke about the possibility of moving some of the Makepeace operations to Canada or even Chile in an attempt to adapt to a warming climate, although he won't discuss global warming.

Approximately 5,500 acres of cranberries are presently cultivated in Chile, Quebec, and British Columbia. Cran Chile currently has 1,200 acres under cultivation in southern Chile, with another 300 acres coming into production. Unlike the peat bogs on Cape Cod, the Chilean berries are grown on uplands and irrigated by streams formed by melting snow from the Andes. Lanco, the town nearest to Cran Chile's cranberry farms, lies halfway between an offshore fault line and a chain of volcanoes that runs the length of Chile. For the growth of cranberries this is a perfect environment; for the residents it may not be so perfect.

The company began in 1992, when Warren Simmons, founder of San Francisco's Pier 39 and Chevy's chain of restaurants, didn't receive the cranberries that went into the cranberry margaritas he featured on his menus. Noting that both Oregon berries and Massachusetts berries are grown at a latitude of plus or minus forty degrees north, he wondered if cranberries couldn't do as well at a latitude of forty degrees south. He was aware that cranberries need sandy soil, fresh water, and a climate that provides the chilling hours required for dormancy. And he found those conditions in Chile's Lakes region, at the latitudes thirty-nine to forty degrees south. An added advantage is that the area around Lanco is almost insect-free, enabling Cran Chile to grow cranberries with few if any applications of pesticides.

Prior to a recent sale of its processing facility, Cran Chile supplied cranberry concentrate and other cranberry products to North America, South America, the European Union, Asia, Africa, and wherever the taste for cranberries had been cultivated and demand existed. The company was vertically integrated, owning its own processing plant and drying and processing equipment to turn harvested berries into sweetened dried berries.

Gordon Swanson Warren Simmon's friend, co-owner, and partner assumed management of the venture, following Warren's death. I ask Gordon why he chose to turn his berries into sweetened dried berries instead of continuing to rely on sales from cranberry concentrate. He replies that converting fresh cranberries to sweetened dried cranberries allows for what he terms "dual usage," meaning that he was still able to extract the juice but also to sell the skins and hulls, previously discarded before the advent of sweetened dried berries.

New developments in the cranberry industry, as in other industries, often succeed in unexpected ways. The move to canned cranberry sauce in the 1900s provided a market for the white and slightly blemished berries. After that, canned cranberry sauce overtook and practically eliminated the market for fresh berries. Today proceeds from the sale of dried cranberries are greater than those from cranberry juice concentrate, originally the primary use for the processed berry. Each of these instances of dual usage has opened up new markets for the cultivated cranberry while spurring growers to increase their yields and, if possible, their acreage under cultivation.

David Mann, who has served on the Ocean Spray board, states that up to a year ago, if a grower wanted to put in a half acre, there'd be a court case over it. "Now," he says, his voice rising, "they [Ocean

Spray] want to plant over five thousand acres. And they've already got an option on the land!"

When I ask where, he replies, "Somewhere in Canada—a former fishing community with good peat soil. For a dollar an acre, or something like that, they can now plant cranberries."

I suggest that having bogs in a more northerly latitude might be a good idea as the planet warms.

"Yeah," David halfheartedly agrees, "and they don't have the problems we have here with insects, regulations . . ."

"In Massachusetts," Marjorie Mann interjects, "we're not even allowed to straighten out a bog where we've been growing cranberries for generations."

"But five thousand acres! We only have a total of ten or twelve thousand acres of bogs in all of Massachusetts." David references the new hybrids his son Keith is planting and their higher yield. "That's the trouble with cranberry growing. It's a perennial crop that takes four or five years to get into production. Meanwhile, you're building up this potential surplus, but by the time the new vines produce berries, you've flooded the market and driven the price down."

In Massachusetts 20 percent of the berries grown in 2012 were Stevens, with a higher yield than the older varieties; in Canada 80 percent of the berries grown are Stevens, and the Canadian government provides incentives for the purchase and planting of the newer hybrids and their even higher yields.

Ocean Spray is beginning to make the public more aware of where its berries come from. Recently, I purchased a bottle of the company's Cranberry Juice Cocktail. The label features the Mays, a family who has been growing cranberries for three generations in British Columbia, Canada. "Meet Our Growers" is the title beside the family photo of the Mays standing in a sea of cranberries.

Charlie Johnson, an independent cranberry grower in Carver and a friend of David Mann's, describes himself as a businessman first and a cranberry grower second. He recently purchased one thousand acres of low-lying land on the Southwest Miramichi River in New Brunswick's Gulf of Saint Lawrence flood plain in Canada. In making the purchase, he is carrying out the advice he gives others to "buy north and buy high." And he isn't referring to price.

Charlie and his son, Van, who works with him, are from a line of proud and independent Finns. They grew up believing in the importance of upholding their good name and their word. Twenty-one years ago the EPA first contended that Charlie had violated the Wetlands Protection Law when he expanded the shape of one of his bogs. At that time Charlie could have paid a fine and continued to grow cranberries. Instead, he has spent over $1 million hiring geologists and lawyers to prove that he did not violate the act. He is now seventy-nine years old and would like to leave a legacy for his grandchildren. For Charlie, New Brunswick is a hedge against the effects of global warming and an opportunity to again grow cranberries. For New Brunswick, Charlie represents jobs for locals who once made a living from the sea. It is the same pattern that took place on Cape Cod in 1812, when schooner captain Cyrus Hall's success with cranberry cultivation began to lure other seamen from ship to shore and from a way of life on the wane to one on the upswing.

Despite the distance and remoteness of their New Brunswick bogs, Charlie and Van welcome the chance to apply some of the newer farming methods. The new bogs will be set up to leach water off through buried tubes, freeing the Johnsons from having to build a network of central ditches. This technique is similar to the one Keith Mann is using on his organic bogs. Charlie and Van contemplate installing a remote-controlled, computerized sprinkler system with pop-up sprinkler heads. A vine setter came with the purchase of the

New Brunswick property, and Van is building a harvester. Thanks to his father's good business sense, Van is one of the younger growers still able to farm cranberries as an independent grower in the United States—and, in his case, also in Canada.

In the United States Charlie and Van grow cranberries on acreage surrounding their homes. Osprey nests dot the property, a pair of bufflehead swim on a flooded bog, swallows fly low seeking insects for dinner, and in the summer months ruby-throated hummingbirds flit from the scrub pine to feeders at the edge of the bogs. Around the corner from where the old screen house used to be, blue heron and great white egrets have built a hatchery. Upward of a hundred heron and egrets sit on nests and nourish their young in the late spring and summer months.

I find Van in the larger of the sheds in Carver, where he builds and repairs the heavy machinery used on the bogs. Most families who grow cranberries have a member of the family or a person on staff who is a skilled machinist. Van is that man for Johnson Cranberries. He has the long, lean build of the long-distance runner he is in his spare time. And he has the competitive nature to be challenged to find a better way. "I just want to do it better than the previous effort," he says. "When dealing with Mother Nature, there's no guarantee for success, but every year I learn a little more. Come this spring, something will happen, and I'll have to devise a new thingamajig to get around it or come up with a new way to handle it. Just the joy of seeing how much you can grow, competing with your neighbors to do your best, trying to see how far you can push yourself, how well you can do, that's all part of growing cranberries. That's why I do it."

Not everyone is convinced that growers have to move to Canada or Chile to continue to farm cranberries. On an existing bog owned by the Manomet Center for Conservation Science, farmer Jen Fried-

rich is conducting a bold experiment with a variety recently hybrid-
ized by Nick Vorsa at the Cranberry Research Station in New Jersey.
Named Scarlet Knight for the Rutgers University mascot, the berry
is expected to ripen and turn scarlet sooner than other varieties, thus
allowing it to be harvested earlier and hopefully making it more
adaptable to changes in the weather pattern. Jen and Dr. Vorsa are
expecting that Scarlet Knight, developed in a state south of Massachusetts with a warmer climate and greater threat of disease, will be
in a better position to thrive in a more northerly climate as the planet
warms.

Up a hill from the farm, Eric Walberg, Manomet's senior scientist
for climate change, sees the greater intensity of storms and their accompanying heavy rainfalls as indicative of one of the most serious
threats to agriculture in portions of the United States. When I met
with him, he printed out several sheets relating to annual temperature and precipitation trends over the past hundred years. For the
period from 1958 to 2007, U.S. annual average precipitation for the
cranberry-growing areas of Massachusetts and New Jersey has increased on an average of 10 to 15 percent. The cranberry-growing
areas of Wisconsin show a similar increase in precipitation, while the
southern coast of Oregon shows a slight decrease in precipitation.

The increases would appear to favor cranberry growth and agriculture in general, at least in the eastern and middle U.S. growing areas. Eric Walberg tells me this is deceptive because it doesn't take the
type and degree of rainfall into consideration. He produces a second
group of printouts showing the percentage change for what is termed
"very heavy precipitation," defined as the heaviest 1 percent of all
rainstorms. From 1958 to 2007 the New England area experienced
a 67 percent increase and the Wisconsin cranberry-growing area a
31 percent increase in very heavy precipitation. Both numbers are

rising. Meanwhile, the average annual temperature has risen three to four degrees during the same period. The Union of Concerned Scientists predicts that if little is done to curb carbon emissions in the next fifty years, Massachusetts will have a climate similar to that of South Carolina, where it is too warm to grow cranberries.

What this means is that the beneficial gentle rains that slowly seep down to the roots of cranberries and other plants are being replaced by heavy precipitation falling on parched ground and never reaching the roots of the plants. At the same time the increase in warmer days when the sun bakes the earth are increasing. Thus the plants dry out and die unless more irrigation is applied by the grower. This scenario places a greater drain on the aquifers, and at some point it can be assumed that demand is likely to outpace the supply of water, the farm's yield will be reduced as a result, and ultimately cultivation will have to stop. One might think that warmer temperatures could lengthen the growing season, but in reality higher temperatures slow plant growth because they speed up soil evaporation and transpiration, the rate at which a plant's leaves lose moisture.

On the bog Jen has tried to create a model for other cranberry farmers that could tip the scale in favor of the berry's ability to continue to be grown in the United States. The farm plan was put together by Linda Rinta at the U.S. Department of Agriculture. Once the plan was in order, Jen applied for and received grants of over $100,000 to rebuild the bog to the highest standards, using ten inches of screened sand over peat, laser leveling of the bog, a pop-up automatic sprinkler system to consume less energy, and the ability to flood the bog in six hours or flash flood it if insects threaten the plants. A tail water system, functioning as a giant rain barrel, will allow water to be recycled or pumped back to the system so as not to add to the drain on an already scarce natural resource.

Over the course of two days in June 2012, Jen and about thirty volunteers hand-planted seventy thousand plugs of Scarlet Knight, one per square foot. Now it's time to watch, protect, and wait, while hoping their experiment is a success. Asked what made her choose this way of life, Jen answers simply that it "gets in your blood when you start cranberryin'"—a response not too different from the answer Henry Hall and the other pioneer cultivators might have given.

 EPILOGUE

The cranberry submits to cultivation, but it retains that savour of wildness that is its by birthright. It cherishes that as a cat its spirit of independence. There are little things as well as great that cannot be tamed, cranberries and cats, as well as winds and waters.

—Cornelius Weygandt, *Down Jersey*, 1940

The sun has begun to set a little later and Venus beams more brightly in the southwest quadrant of the night sky. Before dawn an old gray van carrying three men slowly moves along Route 195 from Fall River, Massachusetts, to the Carver turnoff. At the bottom of the ramp, it heads north to two pudding-stone pillars. The van exits the pavement and bumps along a dirt road through some scrub pines and out into an area of broad, reddish brown squares the color of cranberry leaves in winter. The driver parks the truck. The men clamber out, pick up shovels and wood-tined rakes, then fan out across the expanse of muted red, where they begin raking.

They gather up and remove weeds; old-growth leaves, or "duff"; and loose vines damaged during the harvest. Mud and debris will be cleared out of the ditches to allow the bogs to be flooded. Particular attention is paid to removing the invasive dodder weed, a parasitic plant that saps nourishment from cranberry vines and can destroy an otherwise healthy bog.

As the men work, they laugh and share jokes in their native language. In the sixties and seventies, Mak, Wong, and Ban farmed rice on their own plots of land in Cambodian villages. Forced to flee, the three were relocated to the United States, where they farm the cranberry bogs—still farmers, still gardening, still able to work outside. Finland, Ireland, Cape Verde, Brazil, Mexico, and now Cambodia are all woven into the tapestry that makes up the American cranberry story.

These 250 acres comprise the Federal Furnace Cranberry Company bog managed by Gary Weston. Barry Paquin, whose family bog is just on the other side of town, describes them as "some of the best-managed cranberry bogs on the East Coast." Unlike Cape Cod bogs that were built on peat, these bogs and adjacent bogs were built on iron-rich soil. Shovels, pick axes, rakes, and iron kitchen utensils that sustained this and other agricultural communities in the nineteenth century were fashioned from the bog iron mined here.

Cannonballs for the War of 1812 were produced at a foundry on this property and are believed to have been deployed on the USS *Constitution*, the iconic Old Ironsides. Today the same sand once used for artillery is used to encourage cranberry vines to grow upright toward the sun, where they can flower and produce the berry that provides for Gary's livelihood, plus that of Mak, Wong, Ban, and the bog's full-time workers.

Outside it is thirty-nine degrees Fahrenheit at eleven thirty in the morning. The late-morning sun sits low in the sky, highlighting the sand color of the banks. Then it cedes the sky to flat gray clouds stuck onto a lighter gray background. Beneath the muted light, a backhoe digs down into the irrigation ditches, its mud-filled shovel dripping water and algae from between the tines as it rises from the ditch.

Barry Zonfrelli, Federal Furnace foreman of twenty-eight years, drives up in his pickup truck to deliver a stack of heavy-gauge, woven lattice mats, edged with braided rope, metal rings, and grommets strong enough to hold five hundred pounds of the wet sludge that has accumulated during the summer months. The men carry armloads of these mats across the bog and place them alongside the ditch where Jason, one of the five full-time employees, positions the shovel to drop its next load onto the nearest mat. One of the men promptly laces a corner of the mat through the other three to create a mud-filled pocket.

Previously still air gradually gives way to a barely audible humming sound. As the volume increases, a seemingly fragile arrangement of wire surrounding a plastic bubble emerges from above the pine bluff to the south. The large mechanical "dragonfly" lands at the bog's edge only long enough to disgorge a limber, long-limbed body topped by an oversized helmet. From either side of his head, padded ear protection and speaker phone protrude, giving him the appearance of an insect designed by Disney or the studios of Pixar.

The helicopter lifts off. A metal hook on a cord emerges from the hold to hang suspended a few feet above the ground. Then the dance begins. The hook dangles at a height just low enough for the helmeted man to attach it to a grommet previously threaded through one of the mud-holding pockets. Rising effortlessly, the helicopter heads to a flat, sandy area on the bog's bank. There it turns at just the right angle to swing the bag while disengaging one corner of the mat. Mud sails through the air to a targeted spot of sand. The now-flat mat is swung back to the work area and dropped on the bog close to the irrigation ditch.

In one seamless motion, the helicopter swings over to the next pocket, the ground man connects a corner of the mat, and the sequence continues. When all the filled mud mats have been emptied, and the empty mats returned to be filled again, the helicopter ascends and heads back over the ridge to another bog while the men and the shovel continue to dig and rake in anticipation of the its return. Given the cost of gas, labor, and helicopter rental, cooperation and sharing among bog owners is essential. As is timing.

At the first sustained period of below-freezing temperatures, the bogs will be flooded. The vines will sleep, covered by a layer of ice and crust of snow. From now until early March, the landscape belongs to the animals who still forage for food instead of hibernating or migrating. Otters frolic on the snowy slopes leading from the woods down to the bog; hoof prints of deer are visible in the snow at the line where the trunks of pines meet sand; the path of a muskrat can be discerned leading down to one of the ditches. The repetitious chirp of chickadees, and the raucous caw of the crows is all that's heard in the air's chill. If I stretch my hearing, I can occasionally pick out the mournful sound of two young loons conversing on the pond to the north.

In his office within what may be the last remaining cranberry screening house, Gary Weston speaks of the fragile process of growing cranberries at this time and in this place. "I've seen something that nobody else has ever seen, but development pushes so hard. The land has become so valuable."

Many growers have sold off the upland areas, the hills that provided the sand to nourish growth of the berries. Unearthed stones were used in road building, concrete, and seawalls. The excavated land became foundations for new homes, homes that would attract buyers who weren't necessarily fond of helicopters in their backyards, generators running pump houses during periods of frost, and herbicides or pesticides being applied to nearby acreage. For the grower short-term profit often became a long-term headache.

Clark Griffith's family has been in Carver since 1790. For over half a century Clark has farmed the cranberry bogs his great uncle started in 1902. He relates the story of a neighbor who was bothered by the sound of the generator powering the Griffith irrigation system. "In the middle of the night, he went to my bog, opened the door to my pump house, tore out the wires connecting the generator to the watering system, then went back to bed." The Griffith family's cranberry crop and their income for the year were wiped out. The neighbor may have bought his property for the beauty of the nearby cranberry bog, but he hadn't taken the time to learn what it took to produce the red berries he took for granted.

Years later Clark and his wife, Geraldine, chose to place a conservation restriction on thirty-nine acres of open space around their bogs. Clark continues to grow cranberries on bogs he has protected from an incident similar to the earlier one. The restriction cost him some money in fees, partially offsetting the tax credits he received, but he was able to help other neighbors whose house values increased as a result of living next door to conservation land.

I recently drove down Federal Furnace Road, where I anticipated coming out of the canopy of trees that framed the view of the Piney Woods bog. My trip came to a halt behind three National Grid trucks and a detail of police officers stopping traffic. I looked ahead to see a helmeted National Grid employee in a cherry picker systematically felling the branches with a chain saw. By the time the police motioned for me to move forward, the allée of pine branches leading to the Piney Woods bog was gone.

From 1970 to 2010 the population of Plymouth, Massachusetts, increased by more than 300 percent and the population of Carver by more than 475 percent. The price of land in both towns rose accordingly, as did the temptation to sell. Much of the building was done in areas surrounding cranberry bogs. The new home owners want cable plus fiber optics and a reliable power system. National Grid doesn't want to have to handle complaints from the new homeowners if lines are down because of felled branches in a storm. For a power company, the message is simple: remove the trees. For the older cranberry growers, the answer is simpler: don't sell off the uplands.

I have been fortunate to live in a part of the world where the nutrients in the soil and in the surrounding oceans have provided fertile conditions for the growth of some of the most beneficial and delicious fish, fruit, and berries. Over time the jobs of harvesting those fish, fruit, and berries has drawn men and women from other countries who have enriched our population and provided new recipes for converting our local produce to the table. Together, we have gradually learned to improve the means of harvesting, and we have learned to market and sell the products of our work, thus allowing us to continue to live in this special place.

As more people have prospered, the pressure on our environment has increased, altering the land, air, and soil. And we are beginning to see that it might be possible to lose that which has given us so much. Where Wampanoags walked silently, gathering wild cranberries at the edges of footpaths in the woods; where Clark Griffith's ancestors forged pots, saucepans, skillets, stoves, and wash pans from nearby bog iron; where forty to fifty men, women, and children escaped the cities to stay in bog houses and scooped cranberries using tined wooden scoops; where the caramel-colored sand hills provided nourishment to produce the world's finest cranberries, that's where new homes now stand. As newly developed hybrids have made it possible to grow larger berries and produce higher yields and as the market for cranberries has expanded, the land previously available to grow berries has been turned into starter homes. There's not much left for farming cranberries in that part of the country where cranberry cultivation began.

John Decas describes what he sees as the reasons New England growers sell off their land: "In the East families can't make enough to keep their children in the business; in Wisconsin they have the land to keep their children on the bogs. Two hundred barrels per acre is a good year in the East. The family has a party. A two-hundred-barrel yield in Wisconsin . . . they'd rip up the bog."

When tourists stop to photograph the cranberry harvest at Piney

Woods or to visit Dot and Jack Angley's Flax Pond Farm, they probably aren't aware that the peat bogs where most cranberries are grown are repositories for CO_2, the greenhouse gas believed by many to be responsible for global warming and ultimately for the melting of the Arctic summer sea ice. To many an environmentalist, destroying a bog causes more harm to the balance of nature than destroying the equivalent acreage of a rain forest, because the bog keeps more CO_2 from being released into the atmosphere.

Converting a cranberry bog to a golf course or to starter-home lots not only takes a bog out of production but also may be hastening the migration of the cranberry to cooler climates. Ocean Spray sweetened dried berries will soon be shipped from Chile, thanks to the January 2013 purchase of Cran Chile's cranberry-processing plant. The stated aim is to allow the cranberry cooperative to expand "Ocean Spray's global footprint and establish a Cooperative presence in an important cranberry growing region." The business advantages of the acquisition are many, not the least of them being the emerging Chinese market and the relative tariff advantages of shipping a product from Chile.

Surveying the present situation, Gary Weston muses, "I've been so lucky to be part of growing cranberries in the time and place I did. When cranberries are grown in Canada or another country, but not here, cranberries will still be available, but something uniquely American will be lost."

FOR THE COOK (RECIPES)

We may live without conscience
and live without heart,
We may live without poetry, music
and art;
We may live without friends; we
may live without books;
But civilized man cannot live with-
out cooks.

—Doris M. McCue, *Recipes from*
a Cape Cod Kitchen, 1946

The cranberry first appeared in an American cookbook in 1796, when the author recommended it as "an accompanying cramberry [*sic*] sauce" in her directions for roasting a turkey. The book, *American Cookery* by Amelia Simmons, was the first collection of "hints" for cooking "American" food instead of the English dishes that, up to the book's publication, had provided the only available written means of running a household and preparing food. In her introduction Simmons explains that the book is written for orphans like herself and for those finding themselves "reduced to the necessity of going into families in the line of domestics or taking refuge with their friends or relations and doing those things which are really essential to the perfecting [of] them as good wives, and useful members to society."

Although Stephen Day operated a printing press as early as 1639 in Cambridge, Massachusetts, the only books available for over 150 years on most practical matters such as when to plant corn or how to cook were imported from England with names such as *The English Huswife [sic]: Containing the Inward and Outward Vertues Which Ought to Be in a Compleat Woman* by Gervase Markham or *A True Gentlewoman's Delight, Wherein Is Contained All Manner of Cookery* by Elizabeth Grey, Countess of Kent. Neither these books nor the many other English volumes that followed supplied the recipes that aided in the preparation of the fish, fowl, fruits, and vegetables available in the Americas.

American Cookery filled that void. Unpretentious, it arrived paper covered, only forty-seven pages long, and costing twenty-three shillings, three pence, or about the same as the almanac. For the cook, it was far more useful. Its author chose recipes for the book that showed how to cook the dishes she had grown up eating in America. Cornmeal, for example, was unknown as an ingredient to the British. Amelia Simmons's book listed five recipes calling for cornmeal— three of them the often-eaten "Indian Pudding."

On April 28, 1796, the Connecticut District Court issued a copyright to Amelia Simmons, naming her as author. If her accomplishment in writing the book and getting it published didn't distinguish her enough, her introduction of a chemical leaven made from American potash into the recipes for doughs, or what she termed "pastes," revolutionized baking not only in America but, as word traveled, also in Europe. The baking powder that is an essential in much of today's pastry and bread dough recipes harkens back to the original recipes in *American Cookery*.

Demand for the book prompted the author to publish a second

edition in 1800, where she expanded the pages to sixty-four, added a number of recipes, and left out a few less popular ones. After that, the book was widely reprinted in New Hampshire, Vermont, and New York—many of the reprints unauthorized by the author. In the second edition we find the following recipe:

DRIED APPLE PIE

Take two quarts dried apples, put them into an earthen pot that contains one gallon, fill it with water and set it in a hot oven, adding one handful of cramberries [sic]; after baking one hour fill up the pot again with water; when done and the apple cold; strain it, and add there to the juice of three or four limes, raisins, sugar, orange peel and cinnamon to your taste; lay in paste No. 3.

Later in the book two recipes appear under the heading "Cramberry [sic] Tart." The first simply states, "stewed, strained, and sweetened, put into paste No. 9, add spices till grateful and bake gently." There is an assumption of a certain level of expertise in the reader and cook. A variation of the tart recipe tells the reader that cranberry marmalade may be "laid into paste No. 1, baked gently."

The following cranberry recipes have been chosen by me for reasons similar to those that motivated Amelia Simmons. They taste good. I am not assuming the same level of expertise in the kitchen as she did and have tried to make the recipes as simple as possible. All are either from chefs whose creations I have tasted and liked, or they are directions I have cobbled together to make what I hope will become some of your favorite recipes. I hope you like them and welcome suggestions gleaned from your own variations. Enjoy.

USEFUL AMOUNTS

½ cup raw, whole cranberries equals ¼ cup cooked.
¼ cup cooked cranberries equals one serving.
1 pound raw whole cranberries makes 13½ servings.

🍃 Lentil Salad with Fresh Pea Tendrils

Halfway through the summer, fresh pea tendrils appear at the local organic farm stand. Paired with a lentil salad, they provide a perfect meal on a steamy summer evening. When pea tendrils aren't available, try using baby arugula greens in their place.

LENTIL SALAD

1 carrot, peeled and chopped

1 medium onion, chopped

3 large dried apricot slices

1 pound organic lentils

¼ to ½ pound fresh pea tendrils or baby arugula

2 tablespoons extra-virgin olive oil

2 ham hocks

3½ cups chicken stock (or broth for a stronger flavor)

¾ cup sweetened dried cranberries

1. Peel and chop carrot and onion. Cube dried apricots. Thoroughly rinse lentils in sieve under cold water. Rinse and dry pea tendrils. Set aside.
2. In heavy enamel or stainless steel pan with a cover, heat olive oil. When oil is hot but not smoking, add chopped carrot, onion, and ham hocks. Reduce heat to low. Cover and cook for 5 minutes or until onion is soft but not brown. Remove from pot and reserve.
3. Place washed lentils in same pan used to cook carrot, onion, and ham hocks. Add just enough water to cover lentils and bring to

boil. Drain lentils in sieve and rinse again. Return drained lentils to pan. Add the chicken stock and carrot, onion, dried cranberries, cubed apricots, and ham hocks. Bring to boil over medium high heat. Reduce heat and gently simmer uncovered about 30 minutes or until lentils are cooked but not soft. Remove ham hocks and save for a stew. Drain any excess liquid. Transfer lentil mixture to ceramic bowl large enough to mix lentils with the vinaigrette.

4. Prepare the vinaigrette.

VINAIGRETTE

1 small shallot

½ tablespoon mustard

1 tablespoon sherry vinegar

2 tablespoons hazelnut oil

1 tablespoon extra-virgin olive oil

freshly ground black pepper to taste

salt to taste

chives for garnish

1. Finely chop the shallot. In a small bowl, whisk together mustard and sherry vinegar. While continuing to whisk, slowly pour hazelnut oil and olive oil into bowl. Grind pepper and salt into the vinaigrette. Add minced shallots and continue to whisk until thick and emulsified.

2. Another option is to add all ingredients to a small jar and, with the lid tightly closed, vigorously shake until vinaigrette ingredients are thoroughly blended.

3. Pour vinaigrette over lentil mixture and stir to blend.

4. Place bed of pea tendrils in a large salad bowl. Mound dressed lentils, carrots, onions, cranberries, and apricots in center. Sprinkle chives over salad and place some in a small bowl on the table.

Serves 6 as a main course

🌱 Winter Salad

One of our favorite restaurants is the Terrapin Bistro, in Rhinebeck, New York. On a recent visit, my husband and I marveled over the perfect melding of endive, candied walnuts, apples, and a local blue cheese. We began to imagine how it would taste with a few changes and the addition of cranberries. I hope you enjoy the results.

SALAD

4 medium stalks of Belgian endive

$\frac{1}{3}$ cup candied pecans

½ cup sweetened dried cranberries, preferably organic

4 ounces mild, noncreamy blue cheese such as French Fourme d'Ambert or local artisanal cheese

1. Cut ½ inch off each end of the endives. Discard ends. Remove loose leaves from stalk. When no more leaves fall off, slice another ½ inch off bottom and remove loose outer leaves. Continue until you have all loose leaves. Gather leaves into small bunches and slice across the width into ¼- to ½-inch sections. The goal is to keep the leaves and discard any harder portions of the endive. Place sliced endives in salad bowl.
2. Break up pecans by putting them in a strong plastic bag, then lightly hitting the bag with a wooden mallet until pecan bits are no more than $\frac{1}{3}$ inch in diameter. Place in bowl with endive.
3. Add dried cranberries to bowl.
4. Break up cheese and add to bowl.
5. Cover salad with plastic wrap and refrigerate up to 45 minutes or until ready to serve.

VINAIGRETTE

½ cup balsamic vinegar	1 teaspoon coarse salt
2 tablespoons sugar	2 teaspoons Dijon mustard
½ teaspoon freshly ground black pepper	⅔ cup extra-virgin olive oil

1. To make the vinaigrette, place vinegar and sugar in a nonreactive saucepan and boil down until ⅓ cup remains. Cool.
2. Pour into a glass jar with tight-fitting lid. Add pepper, salt, and mustard. Whisk to blend. Slowly pour olive oil into jar. Screw top of jar on tightly and shake until dressing is emulsified and thoroughly blended.
3. Just before serving, drizzle enough salad dressing over the bowl to barely coat the leaves. Toss.

Serves 4 as a side dish

FOR THE COOK (RECIPES)

🌱 Risotto with Dried Cranberries and Smoked Turkey

In this variation of a classic Italian dish, the tart cranberries are the perfect complement to the smoky flavor of the turkey. The combination calls for a good-quality Italian Parmigiano-Reggiano cheese, available at most grocery stores where cheese is sold. This recipe is a favorite at our home. In the early spring, I pick the new chives in our yard for garnish.

5 cups chicken broth	½ cup dry white wine
5 tablespoons unsalted butter	¾ cup sweetened dried cranberries
3 tablespoons extra-virgin olive oil	2 teaspoons finely chopped fresh sage
6–8 medium mushrooms, stems removed, peeled or wiped dry, then coarsely chopped	¾ pound smoked turkey, sliced ⅜-inch thick.
	½ cup freshly grated Parmesan cheese
1 shallot, minced (approximately two tablespoons)	freshly ground pepper to taste
	salt to taste
1½ cups raw Italian Arborio rice	2 tablespoons of minced chives for garnish

1. In a covered saucepan, bring the broth to a boil. Reduce the heat and allow the broth to barely simmer while you prepare the rest of the dish.
2. Over medium heat, melt three tablespoons of butter in a large, heavy sauté pan. Add one tablespoon of olive oil. When hot, but not bubbling, sauté the mushrooms until soft (6–7 mins.). Remove and keep warm. Add remaining olive oil to the mushroom pan and sauté the minced shallot until translucent but not brown, about 1 minute.

3. Add the rice and stir rapidly for about 2 minutes to coat the rice. Raise heat to medium high. Add the wine and stir for about 1 minute until wine is absorbed.

4. Immediately add ½ cup of the simmering broth to the rice mixture and continue stirring with a wooden spoon. Lower heat so that rice and broth are always gently bubbling, but flame is not high enough to cause the rice to catch on bottom of pan. When broth is absorbed and a wooden spoon stirred along the bottom is slowed by the dry nature of the rice, add another ½ cup. Risotto cannot be abandoned at this point. The cook must continue to stir the rice, making sure to incorporate all the bits of cooked shallots, while adding more broth as necessary.

5. After adding about 3 cups of broth, add the cranberries and sage. Continue to stir mixture and add broth, ¼ cup at a time.

6. After about 20 to 25 minutes of total cooking time, the cranberries should be plump and soft. Continue to stir.

7. Using a sharp knife, cut the smoked turkey into ⅜ x 2 x ⅜-inch strips. Add the strips to the risotto.

8. After about 5 additional minutes, test the rice. It should be creamy and tender but not mushy, gluey, or too hard. You don't have to add all the broth, just enough to allow the rice to reach the proper consistency. Return mushrooms to pan and stir in remaining 2 tablespoons of butter.

9. Add the cheese to the pan and stir to blend. Grind freshly ground pepper on mixture. Stir. Taste before adding additional pepper and salt. Garnish with minced fresh chives and serve.

Serves 4 as a main course

🌱 Cape Blanco Cranberry Sauce

This is an elegant cranberry sauce whose flavor is dependent on the type and quality of the berries and the wine. It is designed to use late-harvested Oregon berries. If using a tarter berry, add an extra half cup of sugar.

8 cups fresh or frozen whole cranberries	1 cup Grand Marnier, or a dry Sauvignon Blanc
2 juice oranges	¼ cup crystallized, sliced ginger, coarsely chopped
2 cups sugar	

1. Wash and pick over berries, discarding any leaves, stems, or blemished berries. Place in a nonreactive soup pot or large sauce pan on stove.
2. Peel oranges, remove seeds, and dice fruit, saving juice.
3. Add orange pieces, juice, sugar, and wine to cranberries. Bring to a full boil over medium high heat.
4. Reduce heat and continue cooking until cranberries pop open and mixture begins to thicken.
5. Add chopped crystalized ginger.
6. Pour into hot, previously sterilized jars and seal. Refrigerate until ready to use. Flavor improves after several days.

Yields about two pints of sauce

🐗 Zucchini Farci

A classic French side dish with an American twist, try it when zucchini are plentiful in the fall and you are looking for a new way to enjoy them.

4 small young zucchini (6 to 7 inches long)

1 shallot, or enough to make 1 tablespoon chopped

2 tablespoons extra-virgin olive oil

1 teaspoon curry powder (or more if you prefer a strong curry flavor)

pinch of salt (optional)

1 cup sweetened dried cranberries

4½ ounces of a creamy goat cheese (artisanal if you can get it)

2 tablespoons heavy cream

1. Preheat oven to 400 degrees Fahrenheit.
2. Cut stems from zucchini. Scoop out center of squash, leaving meat and outer skin. Chop the scooped-out center and save.
3. Using a pot large enough to hold the zucchini halves in one layer, fill half full of water and bring to a boil. Add zucchini and blanch for about 6 minutes or until meat is slightly soft but skins are still firm. Transfer to a bowl of ice water. When cool, remove to paper towels to drain.
4. Finely chop shallot. Warm the olive oil in a small sauté pan on a low temperature. Add chopped shallots. Cover and heat on a low temperature until soft, about 5 minutes. Do not brown. Add chopped squash, curry powder, and, if desired, salt. Cook slowly while stirring for another 5 minutes or until squash is wilted. Transfer to small mixing bowl.
5. While squash is cooking, bring a small saucepan of water to boil. Add cranberries and cook until plumped. Drain and add to squash mixture.

6. Place goat cheese in food processor. Add cream and pulse until smooth.

7. Oil a baking dish large enough to hold the zucchini halves in one layer. Spoon the squash and cranberry mixture evenly into each half. Top with a smooth layer of the goat cheese.

8. Bake in center of oven until cheese is just beginning to turn an amber shade and zucchini is soft. Serve hot from oven or keep warm in a warming oven until ready to serve.

Serves 4 to 8 as a side dish

❤ Avocado Cranberry Bread

The concept of substituting fresh avocado for oils in this bread intrigued me. The results are both moist and delicious, and I hope you enjoy it as much as I do.

2 cups flour

¾ cup sugar

1½ teaspoons baking powder

½ teaspoon baking soda

½ teaspoon salt

½ teaspoon cinnamon

1 egg

1 medium ripe avocado, peeled and mashed to equal ½ cup*

¾ cup buttermilk

¾ cup sweetened dried cranberries

¼ cup pecans, loosely chopped

1½ teaspoons lemon zest

1. Preheat oven to 350 degrees Fahrenheit.
2. Butter a 9 x 5-inch oven-proof glass loaf pan.
3. Put dry ingredients in a large mixing bowl and whisk thoroughly to blend.
4. In smaller bowl, beat egg and add mashed avocado and buttermilk. Whisk to blend.

5. Add the avocado mixture to dry ingredients and blend well.
6. Add cranberries, pecans, and lemon zest. Stir. If bread dough is too thick, use a pastry blender to evenly distribute fruits and nuts.
7. Pour into buttered loaf pan and bake about one hour, or until a toothpick comes out clean.
8. Remove from oven and place pan on a wire rack for 10 to 12 minutes. Then turn out of pan onto rack to cool completely.

Yields one loaf

*The avocado should indent slightly when gently pressed with a finger, but it should not be mushy or brown.

🌱 Cranberry Zucchini Bread

When you find you have more zucchini from your garden than you know what to do with, try making this recipe for your friends and neighbors. They'll all be asking you for the recipe—especially your fellow zucchini farmers.

$\frac{2}{3}$ cup unsalted butter, softened, plus extra for buttering pan

2 cups sugar

1 medium to large zucchini, or two cups when mashed

4 eggs

¼ cup orange juice

$3\frac{1}{3}$ cups flour

1 teaspoon baking powder

2 teaspoons baking soda

1 teaspoon cinnamon, freshly grated if possible

½ teaspoon nutmeg, freshly grated

1½ cups fresh or frozen whole cranberries, sorted and halved

1. Preheat oven to 350 degrees Fahrenheit.
2. Butter and flour two 9 x 5-inch oven-proof glass or five nonstick mini-loaf pans.
3. Using a pastry cutter or fork, thoroughly cream butter and sugar in large mixing bowl.
4. Peel zucchini, thickly slice, and place in a food processor fitted with a metal blade. Process until finely chopped but not mushy. Add to creamed butter and sugar. Incorporate beaten eggs and orange juice. Stir to thoroughly blend.
5. In a separate bowl, combine flour, baking powder, baking soda, cinnamon, and nutmeg. Whisk together to thoroughly blend. Slowly add dry ingredients to zucchini mixture and whisk until thoroughly blended.
6. If halved cranberries aren't available, try placing them in a food processor and quickly pulsing once or until cranberries are sliced but not chopped. Add sliced, washed cranberries to the mixture and stir with a wooden spoon.
7. Divide mixture between 2 loaf pans or 5 mini loaf pans. Bake 1 hour and 15 minutes for loaf pans and 45 minutes for mini loaf pans, or until bread begins to pull away from sides of pan and a toothpick comes out clean when inserted in center.
8. Serve sliced with cream cheese or a creamy goat cheese.

Yields one loaf

🌱 Roasted Brussels Sprouts with Maple Syrup, Bacon, and Cranberries

Brussels sprouts prepared with bits of cranberry are as pleasant to eat as they are to look at. The tart flavor of the berries contrasts perfectly with the heavier flavor of the vegetable. Try this as a colorful alternative to your traditional preparation of brussels sprouts.

2 pounds brussels sprouts or miniature brussels sprouts

¼ cup fresh or frozen whole cranberries

3 tablespoons olive oil

3 tablespoons maple syrup

4 slices of bacon cut into 1-inch sections

freshly ground pepper

salt to taste

1. Preheat oven to 350 degrees Fahrenheit.
2. Cut off base of brussels sprouts and, if large, cut sprouts in half.
3. Wash cranberries and discard any imperfect ones.
4. Mix all ingredients together in mixing bowl.
5. Spread contents of bowl evenly over bottom of oven-proof gratin dish large enough to hold sprouts in one layer.
6. Roast for 35 to 40 minutes or until the bacon is crisp and the sprouts are tender and lightly browned.

Serves 4 to 6 as a side dish

🌱 Spicy Sweet Potatoes and Cranberries

Try this recipe as an option on the all white-and-orange Thanksgiving table.

4 pounds sweet potatoes
(4 to 5 large potatoes)

½ cup water

1 cup fresh or frozen whole cranberries

2 tablespoons butter

2 tablespoons pure maple syrup

½ teaspoon chili powder

2 teaspoons ground cumin

1 teaspoon ground ginger

½ teaspoon freshly ground pepper

½ teaspoon salt

1. Preheat oven to 350 degrees Fahrenheit.
2. Using a fork, prick each sweet potato several times. Roast potatoes until soft, 45 minutes to 1 hour. When done, remove to rack and let cool for 10 minutes.
3. Wash berries and discard any that are blemished. Place water and cranberries in small saucepan. Bring to a boil. Lower heat and cook until cranberries pop open, 15 to 20 minutes. Set aside.
4. When potatoes are cool enough to handle, slip off skins and slice into one inch thick rounds. Transfer to mixing bowl. Add butter and mash with potato masher or fork until only a few lumps remain, depending on your preference.
5. Add maple syrup, chili powder, ground cumin, ginger, pepper, and salt. Stir to combine.
6. Add cranberries and liquid from saucepan.
7. Using a strong wire whisk, incorporate all ingredients without crushing the berries. Remove to warm serving dish and keep warm until ready to serve.

Serves 8 as a side dish

🌱 Cranberry Muffins

Generations of children fondly recall trips to Jordan Marsh, America's first department store, where at Christmas they could visit the Boston store's Enchanted Village of swirling life-size dolls and narrow-gauge trains that puffed real smoke as they emerged from mountain passes. Then they could confide to Santa Claus what they wanted for Christmas—if they were good. Before leaving, they could stop off at the first-floor bakery for a dozen of the best blueberry or cranberry muffins anyone ever tasted. Here is the original recipe, plus a yogurt option.

½ cup butter, at room temperature	2 cups flour
1 cup sugar	½ teaspoon freshly grated nutmeg
2 eggs	1 teaspoon cinnamon
1 teaspoon vanilla	2 teaspoons baking powder
2 cups fresh or frozen whole cranberries	¾ cup milk*

1. Preheat oven to 350 degrees Fahrenheit.
2. Cream butter and sugar in a medium to large mixing bowl.
3. Add eggs one at a time. Add vanilla. Blend well.
4. Gently stir in cranberries.
5. Combine dry ingredients and add alternately with milk (or yogurt) to cranberry mixture. Blend thoroughly.
6. Place muffin papers in muffin tins. Fill ½ to ¾ full.
7. Bake for 30 to 35 minutes until muffins are lightly brown, the berries have opened, and a toothpick comes out clean. Cool on rack before serving.

Makes approximately 18 muffins

*As a healthy option, I substitute 1 cup of plain yogurt for the milk and reduce the butter by ¼ cup.

🌿 Maine Blueberry-Cranberry Pie with Lattice Crust

Maybe you have to be from Maine or Canada to prefer the small low-bush blueberries from those two regions. To my mind, they don't have the slightly tart bite of highbush blueberries, but pair them with cran-berries in a pie . . . now that's ambrosia to the palate and a delight for the eyes. The lattice crust allows the colors to hold their own with the flavor.

PIE CRUST

This is my favorite tried-and-true double-crust recipe, especially for a fruit pie (courtesy of Brooke Dojny, *Dishing Up Maine*).

2½ cups flour

2 teaspoons sugar

1 teaspoon salt

½ cup frozen, commercially packaged lard*

6 tablespoons frozen unsalted butter

6–8 tablespoons ice water

1. In a food processor, combine the flour, sugar, and salt. Pulse until blended.

2. Cut lard and butter into ½-inch bits over the flour. Turn food processor on and off until mixture resembles small pea-size par-ticles. Sprinkle 6 tablespoons of ice water over the mixture. Pro-cess until the dough becomes one ball rolling in the food proces-sor. If the pastry is too dry to form a ball, scatter the additional 2 tablespoons of ice water over the mixture and pulse again. When ball of dough forms, turn off food processor and tip bowl upside down over a lightly floured surface until ball of dough falls out.

3. Cut dough in half and wrap each half in plastic wrap. Flatten to two 5-inch round disks. Refrigerate for at least 30 minutes or up to 2 days (may be frozen for up to 1 month, but must be brought

to refrigerator temperature before rolling out). Remove from the refrigerator. Let sit for 10 minutes.

4. Lightly flour rolling pin and place one disk on floured surface. Working from the center and rolling evenly in all directions, roll out the dough until it is 12 inches in diameter. Fold the dough in half and gently place the fold in center of a 9-inch pie plate. Unfold and allow the pastry to settle into the curve of the pie plate. Prick liberally with fork.

*If packaged lard is not available at your supermarket or grocery store, check online vendors but be prepared to pay more.

PIE FILLING

3 cups small Maine or Canadian blueberries	1 tablespoon freshly squeezed lemon juice
1½ cups fresh or frozen whole cranberries	2½ tablespoons flour
1 cup sugar	1 tablespoon unsalted butter
pinch of salt	

1. Preheat oven to 400 degrees Fahrenheit.
2. Wash cranberries and discard any imperfect ones.
3. Place blueberries, cranberries, sugar, salt, and lemon juice in a large mixing bowl and lightly stir with a wooden spoon until the fruit is evenly coated with sugar.
4. Sprinkle flour over the berry mixture and stir to blend.
5. Spoon pie filling into unbaked pie shell and dot the top with cut-up pieces of butter.
6. Roll out the second ball of pie crust and cut it into parallel strips, ½ to ¾ inches wide. Place one of the longest strips across the center of the pie filling. Fold back from center. Place a second

long strip perpendicular to the first. Unfold the first strip over it at the center of the pie. Place a slightly shorter strip parallel to the second strip and about ¾ inch to its side. Gently lift the first strip from the edge so that the new strip feeds under it. Place a strip parallel to the first strip and ¾ inch away from it. Gently lift the two perpendicular strips, one at a time, and feed the new strip under and over each strip.

7. Continue to place strips ¾ inch away from one another, moving out from center and braiding them over and under one another. When you have reached the edges of the pie, cut off the excess, fold the bottom pastry over the strip ends, and crimp with fingers.

8. If you wish to glaze the upper crust, use a pastry brush to brush a thin coat of milk over the lattice work and crimped edge.

9. Place pie on lowest oven rack and bake for 30 minutes. At this point, the crust should have turned a light golden brown at the edges. Cover with tin foil and reduce the oven temperature to 350 degrees Fahrenheit. Continue baking until the juices bubble up through the spaces between the latticework, about 30 to 35 minutes more. Toward the end of cooking time, take off foil if darker crust is desired. Remove pie from oven and place on a rack to cool before serving.

Yields one pie

FOR THE COOK (RECIPES)

☙ Biscotti Decadence

Tosca, a beautifully appointed restaurant housed in a 1910 granary market south of Boston, serves consistently delicious regional Italian cooking. We love to sit at two of four stools at the cooking bar, where we can watch the preparation. Of course we want to order everything. Recently, we tried a delectable dessert biscotti dipped in warm chocolate, created by Pastry Chef Maria Cavaleri, who kindly shared her recipe.

BISCOTTI

4¼ cups all purpose flour, plus extra for rolling out the pastry

1 teaspoon baking powder

1 teaspoon baking soda

1 teaspoon salt

4 large eggs at room temperature, plus one for egg wash

1 cup sugar

1 cup vegetable oil

1 tablespoon vanilla extract

1½ cups sweetened dried cranberries

1. Preheat oven to 350 degrees Fahrenheit.
2. In medium bowl, mix flour, baking powder, baking soda, and salt. Set aside.
3. In large bowl, using an electric mixer fitted with a paddle attachment, beat eggs with sugar at high speed until light and fluffy. Reduce speed to low and slowly pour oil in a steady stream while continuing to mix. Stir in vanilla until thoroughly combined. A handheld mixer fitted with a whisk may be used up to this point but will require shifting to a wooden spoon for incorporating the flour and berries.

4. Gradually add dry ingredients and stir to mix thoroughly. Fold cranberries into dough.

5. Flour a flat surface or pastry cloth and tip bowl upside down to empty dough onto floured surface.

6. Knead dough for about 30 seconds, pulling it toward you. If dough seems sticky, add more flour and continue kneading.

7. Divide dough into four sections and shape each section into a log. Gently flatten each log until it is rectangular and approximately 1 inch thick.

8. Cut a piece of parchment paper slightly larger than a cookie tin and place on top of a cookie sheet. Place the rectangles on the parchment-lined cookie sheet, approximately 3 inches apart.

9. Whisk together 1 egg and 2 tablespoons of water to make an egg wash. Brush each rectangle with the wash.

10. Move oven rack to middle of oven and bake rectangles for ten minutes. Turn pan and bake for another ten minutes. Remove from oven and cool completely.

11. At this point, rectangles may be wrapped and frozen for up to 3 months. If frozen, bring to room temperature before proceeding.

12. Slice cooked rectangles widthwise into ½- to ¾-inch biscotti shapes. Bake on cookie sheet for 6 minutes on 1 side. Turn and bake for 6 minutes on the other side. Remove from oven and cool on racks before storing in wax-lined cookie tins.

Makes approximately 48 biscotti

MARIA'S CHOCOLATE DIPPING SAUCE

¼ pound (one stick)
unsalted butter

¾ cup dark brown sugar,
packed

¼ cup granulated sugar

½ cup light corn syrup

¼ cup cocoa powder

pinch of salt, if desired

1 cup heavy cream, plus extra
for thinning

10 ounces good quality
bittersweet chocolate*

1 tablespoon vanilla extract

1. Place butter, sugars, and corn syrup in large, heavy-gauge saucepan. Cook over medium high until sugars are dissolved.
2. Whisk in cocoa powder and salt. Bring to a boil.
3. Add 1 cup heavy cream and return to boil. Reduce heat and simmer 5 minutes, stirring often.
4. Remove from heat. Add chocolate and vanilla, whisking until smooth. If sauce seems too thick, whisk in more cream.
5. Pour chocolate sauce into individual preheated ramekins and place beside each dessert plate. Pass the biscotti.

*Makes approximately 2½ cups sauce***

*Always use the best chocolate available. Chocolate quality, like wine, determines the flavor of your cooking.

**Chocolate dipping sauce can be refrigerated and reheated in a microwave or double boiler on simmer.

🍂 Harvard Faculty Club Clafouti

Chef Michael Brentana at the Harvard University Faculty Club created this cranberry *clafouti* to assuage New Englanders of the "cranberry guilt" they sometimes develop when they aren't ordering enough cranberry dishes.

2 eggs

⅓ cup sugar, plus 2 tablespoons

¾ cup milk

1 tablespoon cognac

1½ teaspoons grated orange zest

3 tablespoons flour, plus two tablespoons for baking process

3 tablespoons cornstarch

pinch of salt

1¼ cups fresh or frozen whole cranberries

1 tablespoon powdered sugar, plus extra for dusting

1. Preheat oven to 350 degrees Fahrenheit.
2. Butter a 9-inch pie plate. Set aside.
3. Wash cranberries and discard any that are imperfect.
4. Beat eggs in a medium bowl. Then add ⅓ cup sugar, milk, cognac, and orange zest. Whisk to blend.
5. In a small bowl, mix 3 tablespoons of flour with cornstarch and salt. Gradually add to dry ingredients while whisking to eliminate any lumps.
6. Pour ½ cup of batter into prepared pie plate. Bake for 2 minutes until lightly set on bottom. Remove from oven.
7. Sprinkle with 1 tablespoon sugar. Lay cranberries over sugar. Sprinkle with 1 more tablespoon of sugar. Gently pour remaining batter over all.
8. Bake 20 minutes or until puffy. Sprinkle with a light coating of powdered sugar and serve hot or cold.

Serves 6

🌱 Cranberry-Carrot Cake with Pecan Frosting

When my daughter was a toddler, she refused to eat vegetables. I baked many carrot cakes while hoping that the vitamin A and beta-carotene in the carrots would build up her immune system. The cranberries in this recipe also provide proanthocyanidins (PACs) to help minimize the trips to your family dentist. Coconut oil may reduce anxiety.

CRANBERRY-CARROT CAKE

2 cups unbleached flour

1 teaspoon baking powder

1 teaspoon baking soda

½ teaspoon salt (optional)

4 eggs

1¾ cups unrefined sugar

1 teaspoon vanilla extract
(preferably from Madagascar)

1 cup coconut oil

carrots (2 cups when peeled
and grated)

1 cup sweetened dried
cranberries, chopped

1. Preheat oven to 350 degrees Fahrenheit.
2. Grease a 9-inch springform pan or use a nonstick mini-muffin tin.*
3. In large mixing bowl, whisk together flour, baking powder, baking soda, and salt.
4. Combine eggs, sugar, and vanilla in a separate bowl. Using a whisk or handheld beater, beat until a light creamy color and fluffy. Add the coconut oil and whisk to combine.
5. Add peeled, grated carrots and chopped cranberries to liquid ingredients. Stir.
6. Add cranberry-carrot mixture to dry ingredients and stir well.
7. Pour batter into the prepared pan. Set in middle rack of oven and bake for 1 hour or until cake separates from edge of pan and a toothpick inserted in center comes out clean.
8. Cool in pan. Frost with pecan frosting when cool.

PECAN FROSTING

1 cup chopped pecans, plus extra pecan halves for garnish	4 ounces cream cheese at room temperature
3 tablespoons unsalted butter at room temperature	1½ cups confectioners' sugar
	1 teaspoon vanilla extract

1. In a food processor fitted with blade, process pecans until consistency of granular powder.
2. Cream butter and cream cheese in mixing bowl.
3. Sift confectioners' sugar into bowl in increments, pulsing until fully mixed. If lumpy, keep pulsing until smooth. Stir in vanilla. Sprinkle chopped pecans throughout and gently stir until fully incorporated into frosting.
4. When cake is cool, remove to serving plate and frost top, allowing some to slide down the sides.

Serves 12 as cake or makes approximately 50 mini-cupcakes

*For an afterschool treat, instead of pouring cake batter into a springform pan, fit mini-muffin papers into a greased or nonstick mini-cupcake tin. Bake for 20 minutes or until a toothpick inserted in cupcake center comes out clean. Frost as usual. Decorate each cupcake with pecan half.

🍷 Blancmange with Cranberry Coulis

On our seacoast, natives used to gather a white seaweed that washes up on the beaches in late June. After drying it, they made a powder that was used to gel milk. By adding sugar and flavoring they made blancmange, the delicate custard that has been served in European homes for hundreds of years. This recipe uses readily available powdered gelatin and is served with a cranberry coulis: an elegant dessert.

CUSTARD

2 tablespoons cold water	¾ cup superfine (caster) sugar
1¾ tablespoons unflavored gelatin	½ teaspoon orange flower water
2 cups whole milk	

1. Lightly oil 6 half-cup ramekins. Set aside.
2. Measure water into small bowl and sprinkle gelatin over it. Whisk to blend. Let sit one minute.
3. Meanwhile, pour $\frac{1}{3}$ cup of milk into a small saucepan and bring to boil. Add gelatin and water mixture and stir until gelatin is dissolved. Remove from heat. Add remaining milk, sugar, and orange flower water. Whisk together and place over a large bowl of ice water while stirring constantly until mixture is consistency of thick cream. Remove pan.
4. Divide custard mixture among 6 ramekins.
5. Cover each ramekin with plastic wrap and chill for 4 to 6 hours.

COULIS

½ pound cranberries	2 tablespoons freshly squeezed lemon juice
½ cup superfine (caster) sugar	

1. To prepare coulis, purée cranberries and sugar in a blender or food processor. Strain sauce in fine-mesh strainer. Use a wooden spoon to stir cranberry mixture in strainer until all juice has been extracted.
2. Stir in lemon juice and cool in refrigerator until blancmange is set.

TO SERVE THE BLANCMANGE AND COULIS

1. Unmold blancmange onto individual dessert plates by briefly dipping bottom of each ramekin in warm water, then placing dessert plate on top of ramekin and flipping it upside down, while lightly tapping bottom of ramekin.
2. Pour a thin layer of coulis around, not on, each blancmange form and serve. A mint leaf may be placed in center of each blancmange for garnish, if desired.

Serves 6

♥ Toll House Cookies, Updated

Ruth and Kenneth Wakefield were the venerable owners of the Toll House restaurant, home of the original Toll House cookie. To me, they were my friend Don's parents. Whenever I would go to their home, a plate of freshly baked cookies, often warm from the oven, would be waiting on the kitchen counter. I have replaced the nuts with cranberries, reduced the sugar, and substituted organic sugar and whole wheat flour. I hope "Mama," as Mrs. Wakefield was known by her kitchen staff, would approve.

1 cup unsalted butter, softened	¾ cup whole wheat flour
¾ cup brown sugar	1 teaspoon baking soda
¾ cup organic sugar	½ teaspoon salt, as desired
2 eggs	1¼ cup sliced fresh or frozen cranberries*
1 teaspoon vanilla	¾ cups best-quality dark chocolate bits
1½ cups all-purpose flour	

1. In large bowl, using a pastry blender, mix butter and sugars until creamy and uniformly smooth.
2. Add beaten eggs and vanilla to mixture. Stir to blend thoroughly.
3. In separate bowl, mix flours, baking soda, and salt, taking care to thoroughly incorporate baking soda.
4. Stir dry ingredients into liquid ingredients. Gently mix in cranberries and chocolate.
5. With a teaspoon, pick up a spoonful of dough, lightly form into a ball, and place on flat tray in freezer overnight. Transfer frozen balls to plastic freezer bags. Seal and keep frozen, ready to bake whenever you want a plate of cookies.
6. To bake cookies, preheat oven to 375 degrees Fahrenheit. Move oven rack to third shelf from top. Remove frozen balls from freezer ten minutes ahead of baking. Place cookie balls 2 inches apart on cookie sheet. Bake for 12 to 15 minutes. Enjoy.

Makes 50 to 60 cookies

*If presliced cranberries are not available, you may try slicing your own or use sweetened dried cranberries. If you prefer a sweeter cookie, you may substitute semisweet chocolate bits, as in the original.

🌱 Rhubarb-Cranberry Grunt

Consider this an adaptation of a seventeenth-century English dessert with American ingredients, but better, sort of an Amelia Simmons approach. Rhubarb is a harbinger of spring when it arrives at organic farms. My daughter found the original recipe, minus the cranberries, in England.

1¼ pound rhubarb, cut into 1- to 2-inch pieces	½ pound sugar, plus 3 tablespoons
¾ cup fresh or frozen whole cranberries	8–9 fluid ounces double cream*
2 ounces butter	5 ounces flour

1. Preheat oven to 375 degrees Fahrenheit.
2. Wash rhubarb and cranberries, discarding any imperfect berries.
3. Grease shallow oven-proof glass baking dish, approximately 7 x 11 x 1¾ inches.
4. Scatter washed rhubarb and cranberries evenly over bottom. Dot top of fruit mixture with bits of butter and sprinkle with ½ cup sugar.
5. In small bowl, combine double cream, flour, and 3 tablespoons sugar to make a sticky dough. Press dough over fruit, sugar, and butter in clumps.
6 Bake for 45 minutes. I like to serve this with a pitcher of heavy cream.

Serves 6–8

*Double cream or English clotted cream is often located in the cheese department at many grocery stores.

🍂 Barbara Grygleski's Cranberry-Citrus Juice

When I walked into Barbara Grygleski's Wisconsin kitchen, I was greeted with a warm smile and the welcome aroma of cranberries cooking on the stove—lots of cranberries. An offer of something to drink introduced me to cranberry juice that has me happily making my own from Barbara's recipe. It may inspire you to do the same.

2 pounds fresh or frozen whole cranberries	½ stick cinnamon*
3 juice oranges	2 cups sugar*
3 lemons	7 quarts water

1. Wash cranberries and place in a large, nonreactive kettle or soup pot.
2. Peel oranges and lemons and remove white fiber. Discard pulp, seeds, and peel. Thinly slice the citrus and add to cranberries.
3. Add cinnamon stick, sugar, and water. Stir with a wooden spoon until mixture comes to a full boil. Boil uncovered for about 25 minutes or until berries have popped and released their juices.
4. Remove from stove and cool. Strain through cheesecloth or a fine strainer, pour into glass juice bottles, and refrigerate. Juice may be kept in refrigerator for approximately two weeks.

Yields approximately 2 gallons of juice

*This recipe is for a juice that is not as sweet as the prebottled variety. I prefer it that way, but you may add more or less sugar as your individual taste prefers. I also leave out the cinnamon stick

🌷 Cranberry-Banana Pie

This is the pie that I and other family members have made to bring to every bake sale and potluck or to give to a friend who has needed cheering up over a cup of tea. It is easy, colorful, and delicious—a perfect finale to any meal.

3 ripe bananas

1½ cans whole cranberry sauce

pinch of salt

⅓ cup of water

1½ envelopes of unflavored gelatin

one single-crust pie shell, baked (see the recipe for pie crust or purchase a frozen pie shell and follow directions for cooking)

1. Using a masher and a large nonreactive bowl, incorporate the bananas into the cranberry sauce. Add salt.
2. Heat water to boil and remove from heat. Soak gelatin in water while stirring until it is a clear liquid. Add to cranberry mixture.
3. Spoon cranberry-banana mixture into cooked pastry shell.
4. Chill, but don't freeze.
5. Slice bananas on top and serve*

Yields one 9-inch pie

*My mother whipped half a pint of heavy cream with half a teaspoon of vanilla and decorated the pie with the resulting Crème Chantilly instead of slicing bananas on it.

 ACKNOWLEDGMENTS

Some of the kindest and most gracious people are drawn to a life revolving around cranberries, and it gives me pleasure to be able to say thank you to them for their help in bringing a world they know well to the pages of this book. I hope they are pleased with the results, and I am grateful for their generosity, hospitality, and willingness to explain the various processes that go into growing, harvesting, and processing one of our three native fruit still cultivated.

Two of my earliest cranberry teachers are not included in the book, namely Geoffrey (Jeff) LaFleur, former executive director of the Cape Cod Cranberry Growers' Association, who spent many hours with me explaining the workings of his industry, and Carolyn Gilmore, for many years editor and owner of *Cranberry Magazine* who kindly lent me her disks of back issues. Special thanks to Virginia Valiela and her introduction to a unique partnership between cranberries and refuse—a topic for a different book.

Several people included in the book spent countless hours with me and allowed me to visit their cranberry operations at various seasons when different stages of cranberry growth were taking place. Gary Weston, whose family members have grown cranberries for four generations, Charlie Johnson who introduced me to other growers and a way of life, and whose wife, Jean, fed me many cups of soup and sandwiches in their Carver, Massachusetts, kitchen, and John Decas, who always tried to keep me up to date on the industry and whose depth of knowledge I have only touched upon, deserve extra mention.

This book owes a debt to many scientists who kindly shared their knowledge with me on two subjects critical to the topic. Through my husband Richard O'Connell I am fortunate to have access to a world of men and women who study and monitor various aspects of climate change. Without their input, I would have been limited to whatever I was able to read on the climate change debate. I am also indebted to Eric Walberg and Jen Friedrich at the Manomet Center for Conservation Science, Eric for providing me with climate models and Jen for sharing her attempts to promote cranberry sustainability.

Two scientists who kindly welcomed me into their worlds deserve special mention. Without them, this would be a different and more superficial book. Frank Caruso, plant pathologist formerly with the UMass Cranberry Experiment Station, allowed me complete access to his data of more than thirty years of flowering times and Richard Primack not only welcomed me into his lab, but introduced me to his graduate students at the time and listed me as a co-author on a scientific paper. For a scientist of his stature, this was an especially generous offer and one he could have easily chosen not to make.

Special thanks go to David Mendes for taking the time to make me look more broadly at what might be causing the disappearance of the bees in this country. Without his prodding, this book would have been less balanced in its handling of the subject.

I have benefitted from having access to several libraries where the staff not only allowed me to use their facilities, but also helped me locate information that would have been difficult for me to find on my own. The Arthur and Elizabeth Schlesinger Library at the Radcliffe Institute was an invaluable resource for material on cranberry cooking history and lore. For information on the early immigrant and child labor practices on cranberry bogs, I was fortunate to be given access to the material housed at the Falmouth Historical Society, particularly the Spinner Collection. Some of the most complete archives on material relating to the cranberry industry can be found in the Cranberry Rooms at the public libraries of the towns of Carver and Middleborough, MA. I want to thank the librarians in both of these institutions for their help in locating pertinent historical data. Additional thanks also to the staff of the New Bedford Whaling Museum Research Library and Archives for access to its treasure trove of material and archives on the history of whaling.

A manuscript is only a manuscript until an editor at a publishing house or press decides it has the potential to be polished and released as a book. I am extremely grateful to Phyllis Deutsch, editor in chief at UPNE, for choosing to publish this book and for assigning Susan Silver as my ever-patient and hard-working copy editor. Without Susan's unflagging assistance, this book would still be a jumble of words. I would also like to thank Penny Axelrod and Marjorie Wunsch, two friends who were willing to read the manuscript and, in Marjorie's case, kindly provide editorial guidance.

My deepest thanks and appreciation go to my daughter Lily, a finer mind than I can hope to have, who was constantly a source of encouragement and good sense, and to Richard without whom this would have remained just another book on local history.

NOTES

INTRODUCTION

vii "satellite images": Daniel Schrag, "Science and Advocacy, the Legacy of Silent Spring" (lecture, National Snow and Ice Data Center, Harvard University, Cambridge, MA, September 27, 2012).

viii "strange blight": Rachel Carson, *Silent Spring* (New York: Houghton Mifflin, 1962), 2.

viii comparison of software engineers to farmers: David Streitfeld, "In an Apps Boom, Uncertain Payoff," *New York Times*, November 18, 2012.

x "nowhere else on earth": Tom Larrabee Sr., interview with the author, Nantucket Island, MA, May 23, 2006.

CHAPTER 1. A PERFECT DESIGN

1 "choicest product": Beatrice Ross Buszek, *The Cranberry Connection* (Nova Scotia: Black, 1977), 163.

1 In 1677 New Englanders are reported to have sent "ten barrels of cranberries, two of corn mush, and 1,000 codfish" to King Charles with a letter explaining why the colonies were not bound by the laws of England. See Mark Kurlansky, *Cod* (New York: Walker, 1997), 88.

1 "cure for homesickness": While Benjamin Franklin was in London to plead colonists' cause, he asked his daughter to send cranberries to ward off his feeling of homesickness, as reported in James Trager, *The Food Chronology: A Food Lover's Compendium of Events and Anecdotes from Prehistory to the Present* (New York: Holt, 1995), 169.

2 "served to General Grant's men": Better Homes and Gardens, *Five Seasons Cranberry Book* (Des Moines: Meredith / Ocean Spray Cranberries, 1971), 7.

2 "called it a 'craneberry'": Nancy Cappelloni, *Cranberry Cooking for All Seasons* (New Bedford, MA: Spinner, 2002), 12.

3 "Four or five Portuguese": Ann Sears, "Roderick K. Swift Diary," April 10, 11, 1900, in *Spritsail: A Journal of the History of Falmouth and Vicinity* 19, no. 1 (Winter 2005): 7.

4 For more on the Resettlement and Reintegration Program, see Grant Curtis, *Cambodia Reborn? The Transition to Democracy and Development* (Washington, DC: Brookings Institution Press, 1998), 175.

9 Material on Jan Sipkes Cupido: Joan Kerian, Jeffrey LaFleur, Sheila Lawton, Lydia Mathias, and Irene Sorenson, *All 'Bout Cranberries* (Wareham, MA: Cape Cod Cranberry Growers' Association, 1993), 3.

9 "between the ice age": Carl Waldman, *Atlas of the North American Indian* (New York: Facts on File, 1985), 27.

9 *"sassamanesh"*: Cappelloni, *Cranberry Cooking*, 12.

9 "hearty soup": Carol Wynne (lecture, Lunch and Learn series, Aptucset Trading Post Museum, Plimoth Plantation, Plymouth, MA, November 7, 2013).

9 *"pimmegan"*: James Isham, *1743 Observations on Hudson's Bay and Notes and Observations on a Book Entitled*, A Voyage to Hudson Bay in the Dobbs Galley, 1746–7 (London: Hudson's Bay Record Society, 1949), 155. As documented in Charlotte Erichsen-Brown, *Medicinal and Other Uses of North American Plants: A Historical Survey with Special References to the Eastern Indian Tribes* (New York: Dover, 1979), 206.

9 "ibimi": Buszek, *Cranberry Connection*, 6.

9 "as a poultice": Benjamin Eastwood, *The Cranberry and Its Culture: Complete Manual for the Cultivation of the Cranberry, with a Description of the Best Varieties* (New York: Saxton / Agricultural Book Publishers, 1847), 13.

10 "suffer from 'cancers'": Lydia Maria Francis Child, *The American Frugal Housewife*, 12th ed. (Boston: Carter and Hendee, 1832), 116.

10 "as a diuretic": Cappelloni, *Cranberry Cooking*, 14.

10 "sailed to Boston": John Josselyn, *New-Englands Rarities Discovered* (1672; repr., Bedford, MA: Applewood Books, n.d.), 66.

11 "which grows Cranberries" Merriweather Lewis and William Clark, *The Journals of the Lewis and Clark Expedition*, vol. 6, edited by Gary E. Moulton (Lincoln: University of Nebraska Press, 1990).

11 "for dyeing rugs": Eastwood, *Cranberry and Its Culture*, 15–17.

12 "some cranberries for the sick": Ibid., February 1806 journal entry.

12 "curative powers": Ibid., 13.

12 "often fatal disease": Josselyn, *New-Englands Rarities Discovered*, 65–66.

NOTES

12   "bills of lading": Bills of lading, folder 1, series A S-S2, MS 44, Whaling Museum Research Library and Archives, New Bedford, MA.

13   "letter home": Catherine Parr (Strickland) Trail, "The Backwoods of Canada—Being Letters from the Wife of an Emigrant Officer," 1838, expurgated from Charlotte Erichsen-Brown, *Medicinal and Other Uses*, 207.

13   "swamp land grant": Dave Engel, *Cranmoor: The Cranberry Eldorado* (Rudolph, WI: River City Memoirs, 2004), 9.

13   "Western tribal story": Bill Snyder, interview with the author, Coos Bay, OR, November 9, 2009.

14   "collected plants": Huron Smith, "Ethnobotany of the Forest Potawatomi Indians," *Bulletin Public Museum of Milwaukee* 7 (1933): 32–127.

14   "medicinal practices were sacred": Erichsen-Brown, *Medicinal and Other Uses*, 206.

14   "Any woman": Not all scientists agree with this remedy. A 2011 article in the *New York Times* reports that the ability of cranberries to cure or prevent bladder infections can't be easily proven to work. See Abigail Zuger, MD, "Reputation of a Berry Is Difficult to Confirm," *New York Times*, January 31, 2011.

15   "structure of PACS": Catherine C. Neto, Christian G. Krueger, Toni L. Lamoureux, Miwako Kondo, Abraham J. Vaisberg, Robert A. R. Hurta, Shannon Curtis, Michael D. Matchett, Horace Young, Marva I. Sweeney, and Jess D Reed, "MALDI-TOF MS Characterization of Proanthocyanidins from Cranberry Fruit (*Vaccinium macrocarpon*) That Inhibit Cell Growth and Matrix Metalloprotein Expression In Vitro," *Journal of the Science of Food and Agriculture* 86 (2006): 18–25.

15   "apple juices": Amy. B. Howell, Jess D. Reed, Christian G. Krueger, Renee Winterbottom, David G. Cunningham, and Marge Leahy, "A-Type Cranberry Proanthocyanidins and Uropathogenic Bacterial Anti-adhesion Activity," *Phytochemistry* 66, no. 18 (2005): 2281–91.

15   "Native Americans believed": Constantine Samuel Rafinesque, *Medical Flora or Manual of Medical Botany of the United States*, vol. l. (Philadelphia: Atkinson and Alexander, 1830), as noted in Erichsen-Brown, *Medicinal and Other Uses*, 206.

16   "prohibit bacteria from adhering": Catherine Neto, interview with the author, Dartmouth, MA, March 14, 2007.

16   "bacteria blocker": James A. Greenberg, Sara J. Newmann, and Amy B. Howell, "Consumption of Sweetened Dried Cranberries versus Unsweet-

ened Raisins for Inhibition of Uropathogenic *Escherichia coli* Adhesion in Human Urine: A Pilot Study," *Journal of Alternative and Complementary Medicine* 11 (2005): 875–78.

CHAPTER 2. THE FIRST CULTIVATORS

18 The word "Banks" refers to Georges Bank and Grand Bank, the traditional fishing grounds for Cape Codders.

18 "ordinance in 1773": Bernard T. McGowan, "Financing the Cranberry Crop" (master's thesis, Graduate School of Banking, Rutgers University, June 1952).

19 "Samuel Sonnett": Edward G. Hudson, "Facts Concerning the Story of Cranberries," 1939, archived at Middleborough Public Library, Middleborough, MA.

19 "granted ownership": Henry S. Griffith, *History of the Town of Carver, Massachusetts: A Historical Review, 1637–1910* (New Bedford, MA: Anthony and Sons, Printers, 1913), as referred to in the "Massachusetts Historical Commission's Reconnaissance Survey Report, Carver," 1981, prepared as part of the regional report for southeast Massachusetts, archived at Middleborough Public Library, MA.

19 "Henry Hall": Robert Demanche, "The Early Cultivators," in *Cranberry Harvest: A History of Cranberry Growing in Massachusetts*, ed. Joseph Thomas (New Bedford, MA: Spinner, 1990), 26–27.

20 "Enoch Doane": Charles Nordhoff, "Mehetabel Roger's Cranberry Swamp," *Cape Cod and All Along Shore* (New York: Harper and Brothers, 1868), 179.

21 "most frequently used": Ed Grygleski, interview with the author, Tomah, WI, July 03, 2009.

21 "'tis early, and 'tis black": Stephen Cole and Lindy Gifford, *The Cranberry, Hard Work and Holiday Sauce* (Gardiner, ME: Tilbury, 2009), 25.

21 "Residents": Charles S. Beckwith and Henry B. Weiss, *A Survey of the Cranberry Industry of New Jersey*, circular 45, Department of Agriculture, State of New Jersey, Trenton, February, 1922, 3.

21 "Cranberries followed blueberries": John McPhee, *The Pine Barrens* (New York: Farrar, Straus, and Giroux, 1967), 44.

21 "Around 1835": Joseph J. White, in *Cranberry Culture* (1885; New York:

Orange Judd, 1912), 21, dates the first cultivation in New Jersey to 1845, but doesn't mention Benjamin Thomas. Thus, I stated the earlier date from Beckwith and Weiss, *Cranberry Industry*.

22   "New Jersey produced": McGowan, "Financing the Cranberry Crop," 45.

22   "cranberry fever": A. J. Ryder, *Proceedings of the Eleventh Annual Convention of the American Cranberry Growers' Association*, Camden, NJ, August 28 (Trenton, NJ: MacCrellish and Quigley, 1883), 20.

22   "McFarlin, another cultivator": Robert Demanche, "Beyond Cape Cod," in Thomas, *Cranberry Harvest*, 34.

23   "As true then": "1929 Crop and Livestock Review," March 1930, Massachusetts Department of Agriculture, archived at Middleborough Public Library, MA.

23   "consumption of": Eastwood, *Cranberry and Its Culture*, 72.

24   "date from 1550": Buszek, *Cranberry Connection*, 58.

24   "We have seen": Eastwood, *Cranberry and Its Culture*, 73.

24   "taking the plants": *Naturalist*, 1832, as quoted by Christy Lowrance, "From Swamps to Yards," in Thomas, *Cranberry Harvest*, 14.

CHAPTER 3. PLANTING

28   "cells will be": Keith Mann, interview with the author, Buzzards Bay, MA, March 24, 2009. The term "cell" in this case refers to one plant plus roots, soil, and potting medium.

29   "state average yield": The 2008 Massachusetts average yield has been surpassed in later years as more growers began to plant the higher-yielding hybrids.

30   "pretty expensive": Marjorie Mann, interview with the author, Buzzards Bay, MA, April 20, 2009.

30   "the patriarchs": Thomas, *Cranberry Harvest*, 217.

37   "roots of the vines": White, "Practical Grower," in *Cranberry Culture*, 34.

39   "three types based on shape": Ibid., 55.

CHAPTER 4. IRRIGATION

41   "frost tolerances vary": Carolyn DeMoranville and Hilary Sandler, *Frost Management* (East Wareham, MA: UMass Cranberry Station, 2006–14).

42 "Jarvis Lovell attempted": Cole and Gifford, *Cranberry*, 26.

43 "We're getting regulated": Iain Ward, interview with the author, Lakeville, MA, April 17, 2009.

49 "I stood on it": Andrew Rinta, interview with the author, Wareham, MA, April 23, 2009.

CHAPTER 5. A MATTER OF INSECTS

54 "Mr. Lovell is reported": Cole and Gifford, *Cranberry*, 30.

54 "Professor Louis": Ibid., 30.

55 "Augustus Leland noticed": Ibid., 26.

55 "sodium cyanide": Robert Demanche, "Building a Cranberry Bog," in Thomas, *Cranberry Harvest*, 56–58.

55 For more on Rachel Carson and cranberries, see Cole and Gifford, *Cranberry*, 187–89.

56 "more days working": Monika Schuler, interview with the author, Mattapoisett, MA, October 27, 2006.

56 "first to offer commercial ipm": John Decas, interview with the author, Carver, MA, December 10, 2012.

58 "Food and Drug Administration": Barbara Constable, comp., "Cranberry Scare of 1959," in *A Guide to Historical Holdings in the Eisenhower Library*, March 1994, accessed January 20, 2014, www.eisenhower.archives.gov.

58 "dangers of the weed killer": Kenneth G. Garside, "Why the Excitement over Aminotriazole?," Ocean Spray newsletter, archived at Middleborough Public Library.

59 "$8,000,500": Christy Lowrance, "Part Three, Working Together," in Thomas, *Cranberry Harvest*, 142–43.

60 "moving toward the poles": Daniel P. Bebber, Mark A. T. Ramatowski, and Sara. J. Gurr, "Crop Pests and Pathogens Move Polewards in a Warming World," *Nature* 3 (2013): 985–88.

CHAPTER 6. ORGANIC, SUSTAINABLE, OR ALMOST CONVENTIONAL

62 "all I do": Domingo Fernandes, interview with the author, Carver, MA, June 24, 2010.

63 "Makepeace was often waiting": "Coming to America," *Bog Builder and*

Family Video, with historical perspective by Dr. Marilyn Halter, historian, immigration studies, Boston University, together with Joseph Spinner, Spinner Publications, archived at Carver Public Library, Carver, Massachusetts, 52 min.

63 "child-labor laws prohibited": Sears, "Roderick K. Swift Diary," 12. photo by Lewis Hine, Spinner Collection, Falmouth Historical Society, Falmouth, MA.

63 "Margaret Fernandes": Margaret Fernandes and her late husband, Silvino (Ed) Fernandes, interview with the author, Plymouth, MA, July 12, 2006.

64 "as many as 1,200 children": "Coming to America."

64 "picking season": White, *Cranberry Culture*, 85.

64 "We did everything": Fernandes and Fernandes, interview.

64 "They could be outdoors": Marilyn Halter, "Cape Verdeans Work the Bog," in *Bog Builder*.

65 "They were happy memories": Mary Andrade, "Cape Verdeans Work the Bog," *Bog Builder*.

65 "Absolutely not": Fernandes, interview.

68 "Good for the majority": Ibid.

69 "Farmers, ranchers, and fishermen": Alain Ducasse, *Harvesting Excellence* (New York: Assouline, 2000), 8.

69 "Every place I go": Scott McKenzie, interview with the author, Port Orford, OR, November 08, 2009.

70 "United Nations defines": United Nations, *Report of the World Commission on Environment and Development: Our Common Future*, accessed January 20, 2014, www.un-documents.net.

73 "Nowhere on the American continent": A. E. Bennett, nineteenth-century Wisconsin cranberry grower, quoted in Engel, *Cranmoor*, 7.

73 "begins with the natives": Engel, *Cranmoor*, 9.

74 "Edward Sacket": Ibid., 16.

75 "Peg Leg": Cappelloni, *Cranberry Cooking*, 32.

75 "cranberry separators": Better Homes and Gardens, *Five Seasons Cranberry Book*, 6.

75 "it was very efficient": Nodji Van Wychen, interview with the author, Warrens, WI, August 4, 2009.

CHAPTER 7. POLLINATION

81   "colony collapse disorder": Holley Bishop, *Robbing the Bees* (New York: Free Press, 2005), 133, 271.

81   "three general categories": David Mendes, telephone interview with the author, December 11, 2013.

82   "to be nontoxic": "Declining Populations of Pollinator Bees," Uncategorized (blog), March 23, 2011, accessed January 20, 2014, www.baneof injustice.wordpress.com.

82   "use as a seed treatment": "Clothianidid and CCD: Fact Sheet," Pesticide Action Network, North America / Beyond Pesticides, accessed January 20, 2014, www.panna.org / www.beyondpesticides.org.

82   "EPA is supposed to protect": David Hackerberg, quoted in Ariel Schwartz, "Timeline of a Bee Massacre: EPA Still Allowing Hive-Killing Pesticide," *Fast Company*, December 15, 2010.

83   "not inspecting beekeepers": Susan Kegley, interview with the author, Berkeley, CA, May 14, 2013.

84   "Without bee pollination": Cornell University study, 1999, as reported by Allison Benjamin and Brian McCallum, *A World without Bees* (New York: Pegasus Books, 2009), 4.

84   "on the bog for": Fernandes, interview.

84   "fewer hours of sun": David Abel, "Study Finds Steep Drop in Bay State's Native Birds," *Boston Globe*, September 19, 2011.

85   "sweetener most commonly used": Benjamin and McCallum, *World without Bees*, 253.

85   "imported German Blacks": Peggy M. Baker, "Vignettes from the Rogers Family: A Taste of Honey," *Compact* 32, no. 2 (2011): 6.

85   "4,000 in North America": Kathleen Hatt, "The Buzz on Bees, Help Native Bees Pollinate Your Crops," *Growing: The Business of Fruits, Nuts and Vegetables*, December 2012, A6–A9.

85   "It takes approximately 100,000": Peter Garnham, "Organic Approach, Pollinators," *Horticulture: The Art and Science of Smart Gardening*, June 2009, 15.

86   "patch duty": A. Rinta, interview.

86   "enough native food": Linda Rinta, interview with the author, West Wareham, MA, March 24, 2009.

87   "as much as 96 percent": Sydney A. Cameron, Jeffrey D. Lozier, James P. Strange, Jonathan B. Kock, Nils Cordes, Leellen F. Solter, and Terry L.

Griswold, *Patterns of Widespread Decline in North American Bumblebees*, abstract, proceedings of the National Academy of Sciences (Urbana: University of Illinois, 2010).

87 "diminished by 87 percent": Gary Paul Nabhan, *Growing Food in a Hotter, Drier Land* (White River Junction, VT: Chelsea Green, 2013), 191.

88 "bees for each order": K. Mann, interview.

88 "Senchon petitioned": Hilda Ransome, *The Sacred Bee in Ancient Times and Folklore* (Cambridge MA: Courier Dover, 2004), 27.

89 "best pollinator": Linda Rinta, e-mail to the author, November 27, 2013.

CHAPTER 8. TRICKING THE VINES

91 "Finnish word for swamp": Cappelloni, *Cranberry Cooking*, 30.

91 "look no further": Linda Donaghy, "The Finns," in Thomas, *Cranberry Harvest*, 90.

93 "You measure it": Driver of a sander, conversation with the author, Oiva Hannula bogs, Carver, MA, January 26, 2008.

CHAPTER 9. THE HARVEST, DRY OR WET?

97 "cause less damage": K. Mann, interview.

97 "a real pain": A. Rinta, interview.

101 "are most effective": Mary Puhl, interview with the author, Cape Blanco, OR, November 8, 2009.

101 "allows the cranberry to float": Ron Puhl, interview with the author, Cape Blanco, OR, November 8, 2009.

103 "Regardless of the causes": Kathy Lynn and Ellen Donoghue, *Climate Change and the Coquille Indian Tribe: Planning for the Effects of Climate Change and Reducing Greenhouse Gas Emissions* (Coos Bay, OR: Coquille Indian Tribe, 2010).

106 "our crews are experienced": Dot Angley, interview with the author, Carver, MA, October 6, 2009.

107 "my grandfather's right-hand man": Matt West, interview with the author, Carver, MA, October 6, 2009.

107 "run forty-two acres": Jack Angley, interview with the author, Carver, MA, October 6, 2009.

107 "independent growers received": Michelle L. Johnson, "Cranberry Farmers Struggle as Prices Plummet," *Huffington Post*, May 6, 2013, accessed January 20, 2014, www.huffingtonpost.com.

107 "$63.11 per barrel for A Pool": Malcolm McPhail, "Ocean Spray Growers, Independent Growers, and the Current Surplus of Cranberries," *Coast River Business Journal*, June 6, 2013, accessed January 20, 2014, www.crbizjournal.com.

109 "golf course": Larrabee Sr., interview.

109 "and they will": Tom Larrabee Jr., interview with the author, Nantucket Island, MA, May 23, 2006.

109 "Almost 50 percent of Nantucket": Larrabee Sr., interview.

110 "about twenty dollars a barrel," 2012 Annual Report (Nantucket, MA: Nantucket Conservation Foundation, 2012), 16.

CHAPTER 10. PROCESSING, HANDLING, AND DISTRIBUTION

115 "Decas Cranberries in Carver": Decas, interview.

116 "in dire poverty": Ibid.

117 "existed since 1864": date from Better Homes and Gardens, *Five Seasons Cranberry Book*, 7.

117 "Dear Sir": letter, folder 1, series B, MS 44, Knight Collection, Whaling Museum, New Bedford, MA, 8.

118 For additional material on early cooperatives, see Dan Georgianna, "Co-operation and Competition: The Ocean Spray Story," in Thomas, *Cranberry Harvest*, 154.

119 "Marcus Urann": Thomas, *Cranberry Harvest*, 154–67.

120 "intelligence and business capacity": Booz Allen Hamilton, "Report for the Fiscal Year Ending May 31, 1943," in *Cranberries: The National Cranberry Magazine, 1945–46* (Wareham, MA: Wareham Courier Office), 27, archived at Wetherby Cranberry Library Digital Archives, accessed January 20, 2014, content.wisconsinhistory.org.

120 "paid 100 per cent": United Cape Cod Cranberry Company, prospectus (Boston: Arakelyan, 1909), 10.

121 "not good for us": Decas, interview.

CHAPTER 11. HYBRIDS TAKEN TO A NEW LEVEL

126   "1899 produced": Judge John A. Gaynor, "Low Prices and Possible Combinations," *Milwaukee Sentinel*, quoted in Engel, *Cranmoor*, 91.

126   Material on crop production from Cranberry Forecast, New England Agricultural Statistics, New England Field Service, National Agricultural Statistics Service, United States Department of Agriculture, Concord, NH, August 16, 2012.

127   "Fiorello La Guardia": Ellen Stillman, interview with the author, Hanson, MA, April 21, 2006.

127   "What's Chicken": advertisement in *Family Circle*, February 1949, 84, assembled by Ellen Stillman as part of Ocean Spray's Chicken 'n Cranberry campaign.

127   "close to 30 percent": "US Cranberries Host International Cranberry Harvest Tour in Wisconsin: Leading South Korean Food and Lifestyle Writers to Experience U.S. Cranberries, October 6–9," PRWeb, October 7, 2013, accessed January 20, 2014, www.prweb.com.

128   "50 percent of increased productivity": Edward O. Wilson, *The Diversity of Life* (Cambridge, MA: Belknap, 1992), 301.

128   "they look good": Decas, interview.

129   "having a hard time": Grygleski, interview.

131   "It looks like": Ibid.

133   "plants are hardier": Abbott Lee, interview with the author, Chatsworth, NJ, March 3, 2010.

135   "every two years": Ibid.

136   "crossed Ben Lear with Franklin": Nicholai Vorsa, interview with the author, Chatsworth, NJ, March 3, 2010.

138   "Percentage of foreign sales": Jeffrey LaFleur, Cape Cod Cranberry Growers' Association (oral report, annual meeting, Cape Cod Commercial Cranberry Growers' Association, Plymouth, MA, 2010).

140   "invited to the party": Decas, interview.

140   "never charged for that": Ibid.

141   "600 percent increase": Nicholi Vorsa and Ajay P. Singh, with the collaboration of Laurent Brard, Rakesh K. Singh, and K. S. Satyan (paper presented at the 234th annual meeting, American Chemical Society, Boston, August 21, 2007).

CHAPTER 12. FROM TART TO SWEET

144   "needed more fat": Daniel E. Lieberman, "Evolution's Sweet Tooth," *New York Times*, June 6, 2012, A27.

145   "more than $1.4 billion": Manisha Juthani-Mehta, "Urinary Tract Infections in Elderly Persons," *Geriatric Nephrology Curriculum*, American Society of Nephrology (New Haven, CT: Department of Internal Medicine, Section of Infectious Diseases, Yale University School of Medicine, 2009), 2.

146   "better product": Decas, interview.

146   "indeed it is": Ibid.

147   "ruled in Decas's favor": Ibid.

147   "In the mid '90s": Robert Reiss, "Ocean Spray's Secrets of Co-op Success," *Forbes.com*, September 15, 2010, 2. (All efforts for an interview by the author with Ocean Spray management were turned down over a period of three years. The *Forbes* article was their recommendation and the only input they wished to contribute.)

147   "only about 70 percent": Gordon Swanson, CranChile CEO, telephone interview with the author, April 14, 2011. The percentage of sales generated by Ocean Spray growers has been as high as 90 percent and as low as 53 percent.

149   "low dosage of antibiotics": Marion E. T. McMurdo, Ishbel Argo, Gabby Phillips, Fergus Daley, and Peter Davey, "Cranberry or Trimethoprim for the Prevention of Recurrent Urinary Tract Infections? A Randomized Controlled Trial in Older Women," *Journal of Antimicrobial Chemotherapy* 38, no. 3 (2009): 255–57.

150   "symptoms in children": Pietro Ferrara, Luciana Romaniello, Ottavio Vitelli, Antonio Gatto, Martina Serva, and Luigi Cataldi, "Cranberry Juice for the Prevention of Recurrent Urinary Tract Infections: A Randomized Controlled Trial in Children," *Scandinavian Journal of Urology and Nephrology* 43 (2009): 369–72.

150   "in older men": Ales Vidlar, Jitka Vostalova, Jitka Ulrichova, Vladimir Student, David Stejskal, Richard Reichenback, Jana Vrbkova, Filip Ruzicka, and Vilim Simaneck, "The Effectiveness of Dried Cranberries (*Vaccinium macrocarpon*) in Men with Lower Urinary Tract Symptoms," *British Journal of Nutrition* 104 (2010): 1181–89.

150   "Cardiovascular disease": Diane L. McKay and Jeffrey B. Blumberg,

"Cranberries (*Vaccinium macrocarpon*) and Cardiovascular Disease Risk Factors," *Nutrition Review* 65, no. 11 (2007): 490–502.

150 "phytochemical extracts": Jennifer Sun and Rui Hai Liu, "Cranberry Phytochemical Extracts Induce Cell Arrest and Apoptosis in Human MCF-7 Breast Cancer Cells," *Cancer Letter* 241, no. 1 (2006): 124–34.

151 "field of antiaging": unpublished (as of 2014) research project on cranberries and antiaging by James Joseph and Barbara Shukitt-Hale, Tufts' Jean Mayer USDA Human Nutritional Research Center on Aging, Medford, MA; see "Pick Berries for Head and Heart Health Benefits," *Tufts University Health and Nutrition Letter*, August 2009.

151 "cranberry and stress reduction": Catherine Neto, Marva I. Sweeney-Nixon, Toni L. Lamoureaux, Frankie Solomon, Miwako Kondo, and Shawna L. MacKinnon, "Cranberry Phenolics: Effects on Oxidative Processes, Neuron Cell Death and Tumor Cell Growth" in *Phenolic Compounds in Foods and Natural Health Products*, ACS Symposium Series, vol. 909 (Washington, DC: American Chemical Society, 2005), 271–82.

151 "cognitive impairment": "Research on Cranberries and Health at UMass Dartmouth Guo Research Group," in *Cranberries and Health* (Dartmouth, MA: UMass Dartmouth, 2014), 1.

CHAPTER 13. WARMING OR CHILLING

153 "atmospheric data": Susan Solomon, Dahe Qin, Martin Manning, Zhenlin Chen, Melinda Marquis, Kristen B. Avery, Melinda Tignor, and Henry LeRoy Miller, eds., *Climate Change 2007: In Contribution of Working Group 1 to the Intergovernmental Panel on Climate Change* (Cambridge: Cambridge University Press, 2007).

153 "approximately 0.9 percent per year": "The Heat Is On," *Economist*, October 22–28, 2011, 99.

154 "almost 60 percent of 305": "Birds and Climate Change: On the Move," National Audubon Society, accessed January 20, 2004, www.birds.audubon.org. For additional information about temperature increases and specific declines in bird population, see Trevor Lloyd-Evans, "Banding Summary: Spring 2013 (15 April–15 June at Manomet," Manomet Center for Conservation Science, accessed January 20, 2014, www.manomet.org.

154 "purple finch": For more on the distribution and conservation status of the

purple finch and other birds, see "Birds in Forested Landscapes," a curriculum plan for linking volunteer birders and professional biologists in a study of the habitat requirements of North American birds. Cornell Lab of Ornithology, accessed January 20, 2014, http://birds.cornell.edu.

154 "five hundred million": Gary Paul Nabhan, "Food Chain Restoration in the Face of Climate Change" (lecture, Arnold Arboretum, Boston, MA, November 20, 2013).

154 "*Plant Hardiness Zone Map*," prepared for the USDA by the PRISM (Parameter-Elevation Regressions on Independent Slopes Model) Climate Group, University of Oregon, released in 2012, accessed January 20, 2014, www.planthardiness.ars.usda.gov/.

154 "nature's cycles of flowering": Jennifer Weeks, "Climate Change Comes to Your Backyard," *Daily Climate*, March 23, 2009.

155 "our research": Carolyn DeMoranville, interview with the author, Wareham, MA, 2009.

158 "*Hardiness Zone Map*": *Hardiness Zone Map*, based on data from five thousand National Climactic Data Center cooperative stations across the continental United States, Arbor Day Foundation, 2006, accessed January 20, 2014, www.arborday.org.

158 "1990 *Plant Hardiness Zone Map*": *USDA Plant Hardiness Zone Map* (Washington DC: U.S. Department of Agriculture, 1990).

158 "229 different living plants": Abraham J. Miller-Rushing, Daniel Primack, Richard. B. Primack, Carolyn Imbres, and Peter Del Tredici, "Herbarium Specimens as a Novel Tool for Climate Change Research," *Arnoldia* 63, no. 2 (2004): 26–32.

159 "plants studied in 2002": Daniel Primack, Carolyn Imbres, Richard B. Primack, Abraham J. Miller-Rushing, and Peter Del Tredici, "Herbarium Specimens Demonstrate Earlier Flowering Times in Response to Warming in Boston," *American Journal of Botany* 91 (2004): 1261–64.

159 "relationships of the plant": Camille Parmesan and Gary Yohe, "A Globally Coherent Fingerprint of Climate Change Impacts across Natural Systems," *Nature* 421 (2003): 37–42.

160 "species at the turn of the century": Abraham J. Miller-Rushing and Richard B. Primack, "Global Warming and Flowering Times in Thoreau's Concord: A Community Perspective," *Ecology* 89 (2008): 332–41.

161 "results of the study": Elizabeth R. Elwood, Susan R. Playfair, Caroline A. Polgar, and Richard B. Primack, "Cranberry Flowering Times and Cli-

mate Change in Southern Massachusetts," *International Journal of Biometeorology* 58 (2013): 1–5. doi:10.1007/s00484–013–0719-y.

162 "have also been migrating": Nicholas S. Fabina, Karen C. Abbott, and R. Tucker Gilman, "Sensitivity of Plant-Pollinator-Herbivore Communities to Changes in Phenology," *Ecology* 221, no. 3 (2010): 453–58.

163 "chill-factor temperatures": Union of Concerned Scientists, "News Reports Suggest Global Warming Is Hurting Coffee Production; How Could It Affect Other Crops?" *Earthwise* 14, no. 2 (2012): 3.

CHAPTER 14. SOMETHING GAINED, SOMETHING LOST

164 "We're having warmer springs": Michael Hogan, quoted in Sarah Gardner, "Marketplace," *National Public Radio*, November 19, 2012.

165 "1,200 acres under cultivation": Swanson, interview.

165 "grown on uplands": Elden J. Stang, "The Emerging Cranberry Industry in Chile" (paper presented at the Sixth International Symposium on Vaccinium Culture, International Society for Horticultural Science, 1996).

165 "Warren Simmons, founder": Rhonda Gordon, "Cranberries," *Emporia (KS) Gazette*, December 8, 2010.

167 "option on the land": D. Mann, interview.

167 "straighten out a bog": Marjorie Mann, interview with the author, Buzzards Bay, MA, July 6, 2009.

168 "buy north and buy high": Charles Johnson, interview with the author, Carver, MA, June 27, 2008.

169 "That's why I do it": Van Johnson, interview with the author, Carver, MA, June 27, 2008.

170 "U.S. annual average precipitation": Thomas R. Karl, Jerry M. Melillo, and Thomas C. Peterson, eds., *Global Climate Change Impacts in the United States* (New York: Cambridge University Press, 2009), map, page 30.

170 "heaviest 1 percent": Ibid., page 32.

171 "temperatures slow plant growth": Andrew T. Guzman, *Overheated: The Human Cost of Climate Change* (New York: Oxford University Press, 2013), 129.

EPILOGUE

174   "best-managed cranberry bogs": Barry Paquin, interview with the author, Carver, MA, May 21, 2009.

176   "development pushes": Gary Weston, interview with the author, Carver, MA, September 10, 2004.

176   "Carver since 1790": Daniel S. Levy, "Waiting for the Frost in Cranberry Land," *Yankee*, November 1994, 92–96.

176   "middle of the night": Clark Griffith, interview with the author, Carver, MA, January 9, 2007.

177   "conservation restriction": Robert Knox, "Tax Benefits May Spur More Landowners to Conserve," *Boston Globe*, March 3, 2011.

177   "the population of": Statistics supplied by the town clerks of Plymouth, MA, and Carver, MA, respectively.

178   "rip up the bog": Decas, interview.

179   "shipped from Chile": "Ocean Spray to Acquire Cran Chile," *United Cranberry* (blog), January 9, 2013, accessed January 20, 2014, www.blogunitedcranberry.wordpress.com.

FOR THE COOK (RECIPES)

181   "American cookbook": Amelia Simmons, *American Cookery; or, the Art of Dressing Viands, Fish, Poultry and Vegetables and the Best Modes of Making Puff-Pastes, Pies, Tarts, Puddings, Custards and Pastries and All Kinds of Cakes from the Imperial Plumb to Plain Cake, Adapted to This Country and All Grades of Life* (Hartford, CT: Hudson and Goodwin, 1796).

182   Gervase Markham, *The English Huswife: Containing the Inward and Outward Vertues Which Ought to Be in a Compleat Woman* (London: Jackson, 1615), 128.

182   Elizabeth Grey, Countess of Kent, *A True Gentlewoman's Delight, Wherein Is Contained All Manner of Cookery* (Padstow, Cornwall: TJ International, 1615), 128.

182   "chemical leaven": Mary Tolford Wilson, foreword to Simmons, *American Cookery* (1796; New York: Oxford University Press, 1958), xiii–xv.

183   "Dried Apple Pie": Amelia Simmons, *American Cookery*, 2nd ed. (Albany, NY: Webster, 1796), 26–27.

183   "laid into paste": Ibid., 29, 30.

195   "Roasted Brussels Sprouts": The inspiration for this recipe comes from a newspaper clipping I saved from our morning paper. See Catherine Smart, "Roasted Brussels Sprouts with Bacon and Maple Syrup," *Boston Globe*, November 25, 2009.

198   "Pie Crust": Brooke Dojny, "Old-Fashioned Lard Crust," reprinted with permission from *Dishing Up Maine* (North Adams, MA: Storey, 2006), 236.

INDEX